Thermal Deconstruction of Biomass 2017

Topical Conference at the 2017 AIChE Annual Meeting

Minneapolis, Minnesota, USA
29 October – 3 November 2017

ISBN: 978-1-5108-5789-6

Printed from e-media with permission by:

Curran Associates, Inc.
57 Morehouse Lane
Red Hook, NY 12571

Some format issues inherent in the e-media version may also appear in this print version.

Printed by Curran Associates, Inc. (2018)

For permission requests, please contact AIChE
at the address below.

AIChE
120 Wall Street, FL 23
New York, NY 10005-4020

Phone: (800) 242-4363
Fax: (203) 775-5177

www.aiche.org

Additional copies of this publication are available from:

Curran Associates, Inc.
57 Morehouse Lane
Red Hook, NY 12571 USA
Phone: 845-758-0400
Fax: 845-758-2633
Email: curran@proceedings.com
Web: www.proceedings.com

TABLE OF CONTENTS

Some abstracts in this book refer to images that were not made available by the publisher

503581 Thermal Deconstruction on Lignocellulosic Biomass

Abstract:

Deconstruction of lignocellulosic biomass into fermentable sugars is among the major challenges in producing cellulosic biofuels and biobased products. The conventional approach to biomass deconstruction uses acid pretreatments to open up plant cell walls, making cellulose more susceptible to acid or enzymatic hydrolysis. Thermal deconstruction, on the other hand, uses heat to liberate plant polysaccharides as anhydrosugars and lignin as phenolic monomers and oligomers. The two major thermal deconstruction approaches are pyrolysis, which occurs in a gaseous environment to produce vapor products, and solvent processing, which occurs in a solvent environment to produce solubilized products. In both cases, plant polysaccharides can be directly solubilized without acid or enzymatic hydrolysis.

Thermal deconstruction of biomass, despite being the foundational phenomenon of thermochemical processing of biomass, is poorly understood. When biopolymers are heated to temperatures in the range of 200-600°C, they undergo a number of chemical transformations, the most prominent including chain breaking, ring fragmentation, dehydration, decarbonylation, and decarboxylation. The relative contribution of each to products, whether solid, liquid, or gas, depends on heating rate, reaction environment (solvents or gases), temperature, residence time of reactants and products in the reactor, and the presence of catalysts. We are investigating fundamental and applied problems in thermal deconstruction using both pyrolysis and solvent liquefaction. We have been able to dramatically increase yields of anhydro monosaccharides from pyrolysis of lignocellulose, which is the basis of commercial demonstration project with Easy Energy Systems. We have found that two-step solvent liquefaction with acid catalyst can substantially increase the rate of formation and yield of solubilized carbohydrates. We are also investigating pyrolysis under very limited time regimes to produce anhydro-cellodextrins, which may prove to be directly fermentable by yeast that produce extracellular enzymes.

503741 Overcoming Challenges of Integrating Pretreatment with Enzymatic Saccharification and Microbial Fermentation Using Ionic Liquids (ILs) or Bionic Liquids (BILs)

Abstract:

Ionic liquids (ILs), solvents composed entirely of paired ions, have been used in a wide variety of process chemistry and renewable energy applications. Imidazolium-based ILs show remarkable abilities to dissolve biomass, and are thus an ideal media for biomass pretreatment and depolymerization. Although very efficient, imidazolium cations are currently expensive and therefore their large scale use and industrial deployment, e.g. in biorefineries, is limited. In an attempt to replace imidazolium-based ILs with ILs derived from renewable sources that retain their efficiency for biomass pretreatment, we have developed a class of biocompatible ILs derived from choline and amino acids. The lower microbial and enzyme toxicity of these recently-developed renewable ionic liquids (ILs), or bionic liquids (BILs), helps overcome the challenges associated with the integration of pretreatment with enzymatic saccharification and microbial fermentation. We increased biomass digestibility at >30 wt% loading by understanding the relationship between ionic liquid and biomass loading, yielding 41.1 g/L of ethanol (equivalent to an overall yield of 74.8% on glucose basis) using an integrated one-pot fed-batch system. Our technoeconomic analysis indicates that the optimized one-pot configuration provides significant economic and environmental benefits for cellulosic biorefineries by reducing the amount of ionic liquid required by ~90% and pretreatment related water inputs and wastewater generation by ~85%. In turn, these improvements can reduce net electricity use, greenhouse gas-intensive chemical inputs for wastewater treatment, and waste generation. The result is an overall 40% reduction in the cost of cellulosic ethanol produced and a reduction in local burdens on water resources and waste management infrastructure.

503124 Biomass Deconstruction Towards Total Carbon Utilization for Future Biorefineries

Abstract:

Lignocellulosic biomass is the most abundant source of organic carbon on Earth and presents the only option with the potential to economically and sustainably replace fossil resources for large-scale production of renewable chemicals, materials, and liquid fuels. However, past methods to achieve the facile deconstruction of biomass has been challenged by processing difficulties, low product recovery, and excessive materials and energy costs. We highlight here the recent developments of novel CELF (Co-solvent Enhanced Lignocellulosic Fractionation) technology to enable more economically feasible deconstruction and conversion of biomass in future biorefineries. At lower reaction severities, we have recently applied CELF to integrate with biological fermentations achieving complete solubilization of solids without added enzymes in just two days culture. In another configuration, CELF paired with high solids ethanol fermentation achieved equivalent performance as compared to pure sugar fermentations. At higher reaction severities, a pathway to high yield co-production of methylated furans from both furfural and HMF was developed using a supported non-noble catalyst. We also demonstrate how atomic-level interactions simulated by supercomputers could help understand key chemical mechanisms responsible for biomass breakdown to help improve design of bulk-level processes that can be subsequently integrated with downstream conversion. The central theme of this discussion will be towards maximizing total carbon utilization of biomass using CELF and a special focus on lignin valorization.

495202 Lignocellulosic Biomass Fractionation and Upgrading Strategies Using Gamma Valerolactone As Solvent

Abstract:

The first step toward a successful biorefinery is the effective deconstruction of lignocellulosic biomass into its main components, cellulose, hemicellulose and lignin to maximize the value of each of those fractions.

GVL fractionation technology is a promising alternative to achieve that objective. Because hydrolysis (100x) and dehydration (30x) reaction rates are faster in GVL than in water, (*1*, *2*) the biomass fractionation step can be carried out at mild conditions (time, temperature, acid concentration, and pressure) producing a high purity solid cellulose stream, and a concentrated liquid hemicellulose and lignin without degrading or diminishing the value of any of the fractions. In addition, the higher solubility of lignin in GVL allows the use of high biomass loading (>30%) reducing the solvent utilization. Solvent recovery is facilitated by the high chemical stability of the GVL, and the fact that GVL can be easily produced from biomass, (*3*) helps to improve the process economics.

The cellulose produced by this technology has high purity and can be used as dissolving pulp, converted into fermentable sugars, ethanol or specialty chemicals such as levulinic acid and HMF. The mild processing conditions used during the biomass fractionation allow to retain the native structure of the lignin that can be isolated at high purity and used to produce carbon foams or battery anodes. Finally, the hemicellulose fraction can be converted into furfural, a specialty chemical, at high yields and without intermediate separations. This technology maximizes the conversion of lignocellulosic biomass into high value products (up to 80% of the biomass to useful products) producing a revenue over $500 per MT of dry biomass which makes the technology economically competitive. (*4*)

1. M. A. Mellmer, C. Sener, J. M. R. Gallo, J. S. Luterbacher, D. M. Alonso, J. A. Dumesic, Solvent Effects in Acid-Catalyzed Biomass Conversion Reactions. *Angew. Chem. Int. Ed.* **53**, 11872â€“11875 (2014).

2. M. A. Mellmer, D. M. Alonso, J. S. Luterbacher, J. M. R. Gallo, J. A. Dumesic, Effects of Î³-valerolactone in hydrolysis of lignocellulosic biomass to monosaccharides. *Green Chem.* **16**, 4659â€“4662 (2014).

3. D. M. Alonso, S. G. Wettstein, J. A. Dumesic, Gamma-valerolactone, a sustainable platform molecule derived from lignocellulosic biomass. *Green Chem.* **15**, 584 (2013).

4. Alonso et al., Increasing the revenue from lignocellulosic biomass: Maximizing feedstock utilization. Submitted to Science Advances (2017)

502724 On the Synergy between Non-Thermal and Thermal Catalytic Deconstruction of Waste Lignocellulosic Materials

Abstract:

In the recent years, lignin valorization has gained upward momentum owing to advances in both plant bioengineering and catalytic processing of lignin. In this new horizon, catalysis is now applied to the 'pulping process' itself, creating efficient methods for lignocellulose fractionation or deconstruction (here referred to as *Catalytic Upstream Biorefining* or 'CUB'). These processes render, together with delignified pulps, lignin streams of low molecular weight (M_w) and low molecular diversity. Recently, we introduced a CUB process based on Early-stage Catalytic Conversion of Lignin (ECCL) through H-transfer reactions catalysed by Raney Ni. This approach renders a lignin stream obtained as a viscous oil, comprising up to 60 wt% monophenolic compounds (M_w <250 Da). The remaining oil fraction (40%) is mainly composed of lignin oligomers, and as minor products, holocellulose-derived polyols and lignin-derived species of high M_w(0.25-2 kDa). Simultaneously, the process yields a holocellulose pulp with low content of residual lignin (<5%).

In this contribution, the contribution of non-thermal catalytic biomass deconstruction to thermal biorefining will be presented, outlining future perspectives in this emerging field of research.

496596 Stabilization with Aldehydes for the High Yield Production of Targeted Monomer Mixtures from Lignin during Integrated Biomass Depolymerization

Abstract:

Lignin is the fraction of lignocellulosic biomass with the highest energy density and is one of the few natural large-scale sources of aromatic molecules. However, the lack of practical depolymerization processes that can produce a limited number of lignin monomers at high yields has limited the options for upgrading lignin. The development of these processes has been restricted by the presence of inter-unit carbon-carbon bonds within native lignin, and further by formation of such linkages during lignin extraction. We report that adding aldehydes and forming their associated protection group during biomass pretreatment produces a soluble lignin fraction that can be converted to guaiacyl and syringyl monomers at near-theoretical yields during subsequent hydrogenolysis. Up to a 48% yield can be achieved from wild type wood while yields as high as 78% can be achieved from high-syringyl transgenic poplar with low levels of carbon-carbon linked units. Furthermore, the nature of the aldehyde and upgrading conditions can be used to reduce the distribution of monomers. The resulting monomer yields were 3-7 times greater than those obtained without using aldehydes as protection groups, which prevented lignin condensation primarily by forming 1,3-dioxane structures with lignin side-chain hydroxyl groups. By depolymerizing cellulose, hemicelluloses, and lignin separately, overall monomer yields between 76 and 90% were achieved for these three major biomass fractions. Recovery of an isolated lignin fraction that is soluble in an organic solvent and can produce high yields of a limited number of monomers could facilitate continuous lignin processing and the development of heterogeneous catalysts for lignin conversion.

Molecular modeling of pyrolysis

Phil Westmoreland,
North Carolina State University

NC STATE UNIVERSITY

Pyrolyze woody biomass, knowing molecular steps.

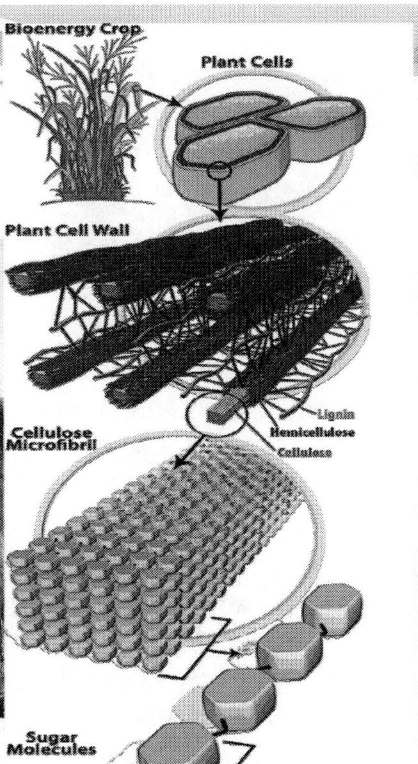

Diagram: DOE/ SC-011.

NC STATE UNIVERSITY

Lumped? Semi-lumped; Molecular.

- Modeling pyrolysis of lignocellulosic biomass with a network of true, elementary reactions is now a plausible goal, not a quixotic vision.
- Biomass contains a great diversity of biopolymers, other species, structures, and biopolymers, including ions.
- For lignocellulosic biomass, deconstruct kinetics as being dominated by cellulose, hemicellulose, and lignin, the three major cell-wall components.
- We can recognize chemical structures and features of kinetics that distinguish these three fractions.

NC STATE UNIVERSITY

In summary for cellulose:

- Polysaccharide of β-D-glucose, linked by 1,4-glycosidic bonds.
  - When aligned, a large number of hydrogen bonds interlink the chains in an orderly fashion, allowing strong crystallinity and microfibril formation.
  - In pure cellulose, the material become amorphous by thermal disruption (a state previously inferred as "active cellulose").
- Then pericyclic reactions cleave the chain to a cello-*n*-san and a shortened cellulose oligomer, ultimately yielding levoglucosan.
  - The pericyclic reactions may be unimolecular steps or OH-catalyzed, bimolecular reactions.
  - When ions are present, chain components are chelated, aiding char.

NC STATE UNIVERSITY

Near-simultaneous 2012 convergence.

- Use of computational quantum chemistry to discover feasible reactions to break down cellulose by concerted, pericyclic reactions – not radical or ionic.

Assary and Curtiss, *CatChemComm* **4**:2 (2012) 200-205: Cellobiose to levoglucosan etc. by concerted rxns, uni & bimolecular

Mayes and Broadbelt, *J. Phys. Chem. A*, 116:26 (2012) 7098-7106: Cellobiose to levoglucosan etc. by unimolecular concerted rxn (C6)

Agarwal, Dauenhauer, Huber, Auerbach, *J. Am. Chem. Soc.* 2012, 134 (2012) 14958−14972: Cellobiose rxn trajectories by CPMD

Seshadri and Westmoreland, *J. Phys. Chem. A*, **116**:49 (2012) 11997-20013: Glucose/cellulose to levoglucosan etc. by concerted rxns, uni & bimolecular

Later support: Zhou, Nolte, Mayes, Shanks, Broadbelt, "Experimental and Mechanistic Modeling of Fast Pyrolysis of Neat Glucose-based Carbohydrates. Parts I. and II." *Ind. Eng. Chem. Res.* 53 (34), 2014, 13274–13289.

NC STATE UNIVERSITY

Crystallinity and rigidity is due to flatness allowed by β ring linkages, to symmetry, and to H-bonds between chains.

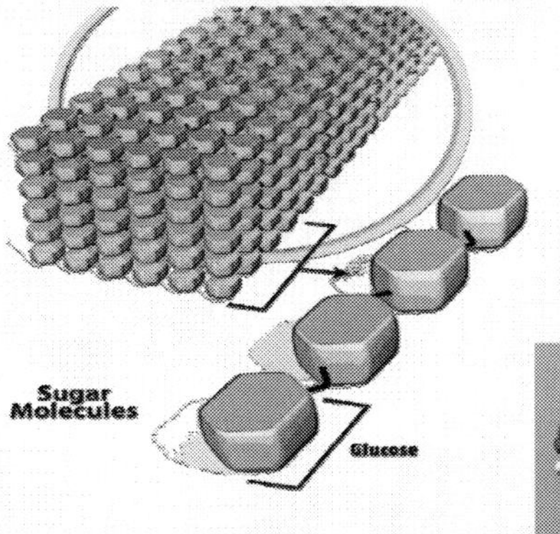

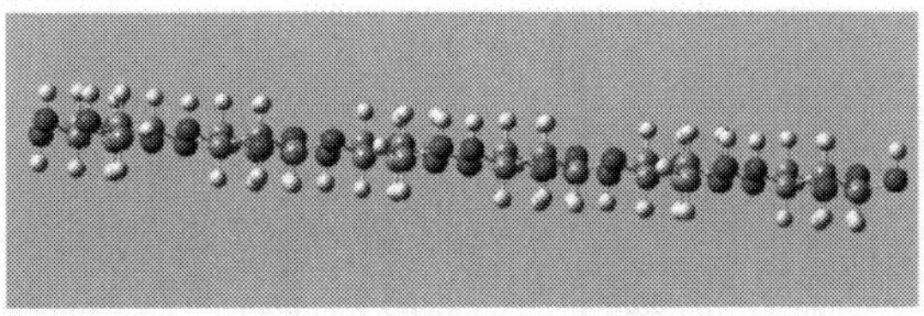

NC STATE UNIVERSITY

Cellulose: Polymer of β-glucose (β-D-glucopyranose).

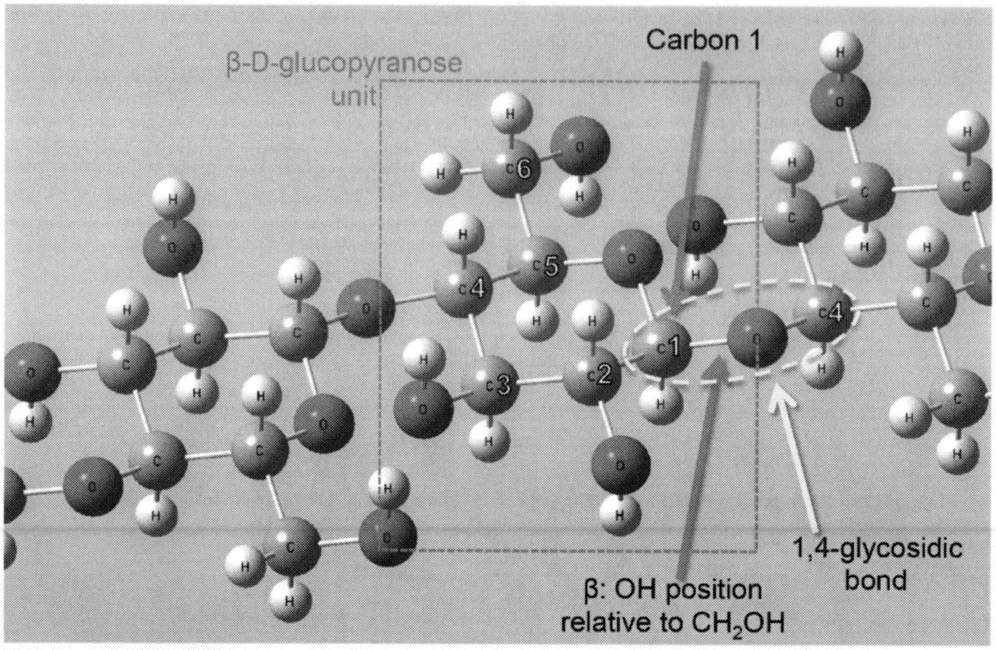

NC STATE UNIVERSITY

Levoglucosan seems to be a key.

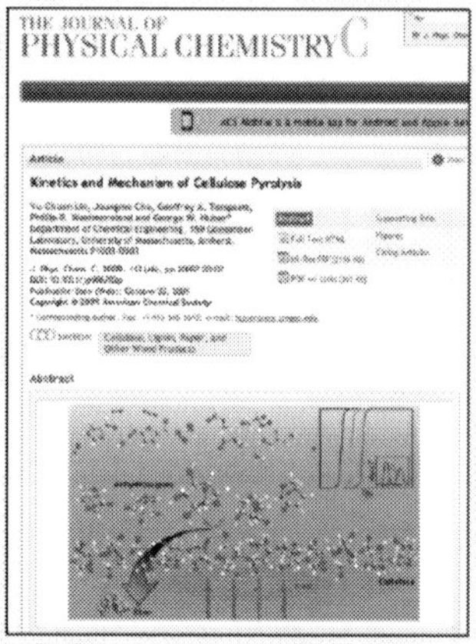
THE JOURNAL OF PHYSICAL CHEMISTRY C

Kinetics and Mechanism of Cellulose Pyrolysis

Abstract

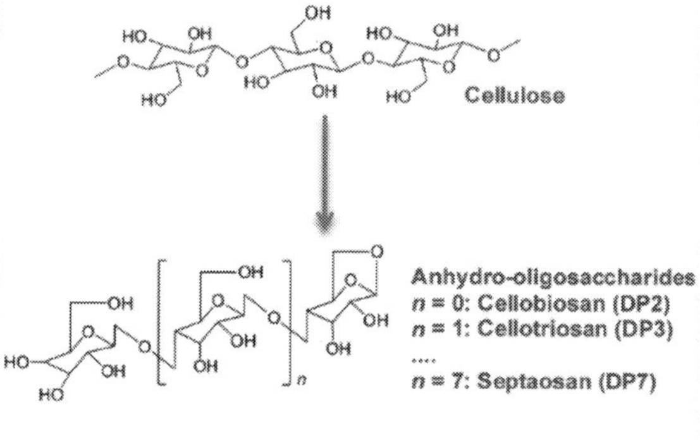

- Our flash pyrolysis with liquid chromatography / MS revealed cello-*n*-san's early formation.
- **Hypothesis: The polymer scission step is pericyclic, breaking the glycosidic bond and forming the bicyclic LGA group.**

NC STATE UNIVERSITY

We probed it with data and computational chemistry, identifying reactions all to be pericyclic.

Seshadri and Westmoreland *J. Phys. Chem A* (2012); Seshadri and Westmoreland, *Catalysis Today* (2016).

- CBS-QB3 (with Gaussian 09) to optimize reactants, transition state, and products, confirming with Intrinsic Reaction Coordinate calcs.

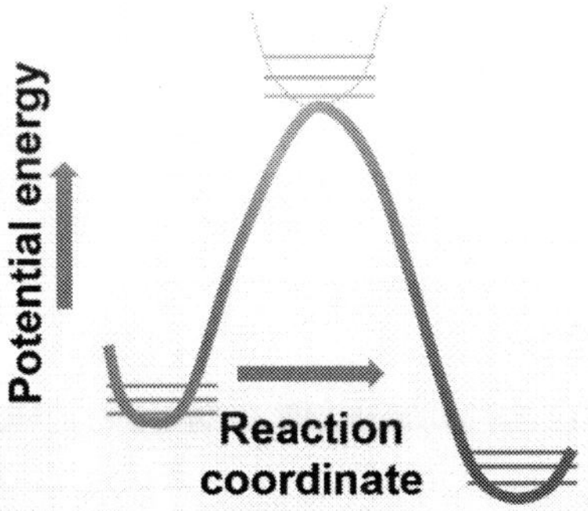

- Solvent effects? Tested explicit H_2Os and Polarizable Continuum Model; no effect.

Found pericyclic reactions (aromatic TSs).

- 6-electron ring has Hückel aromaticity; 4 w/ twist = Møbius.
- Classic example is Diels-Alder reaction of ethylene plus 1,3-butadiene ⇔ cyclohexene:

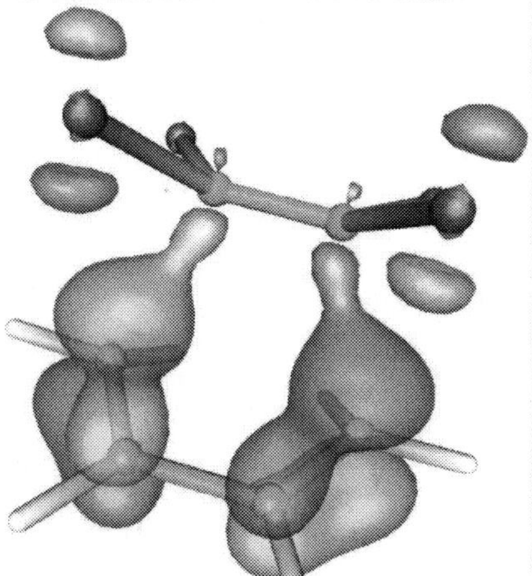

Use glucose as a model compound.

- Is that okay? It gives mix of LGA/AGF and furan-type species ().
- From their data, Sanders et al. (2003) proposed a network of reactions:
  - α- and β interconvertible; can form linear D-glucose.
  - α- and β could form the anhydrosugars levoglucosan and AGF.
  - D-glucose would mainly form furans and furfurals.

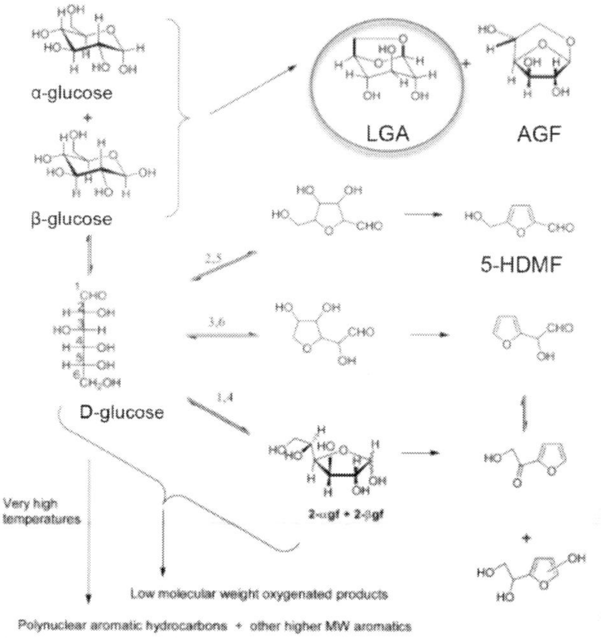

NC STATE UNIVERSITY

Key insights appear with glucose isomerizations.

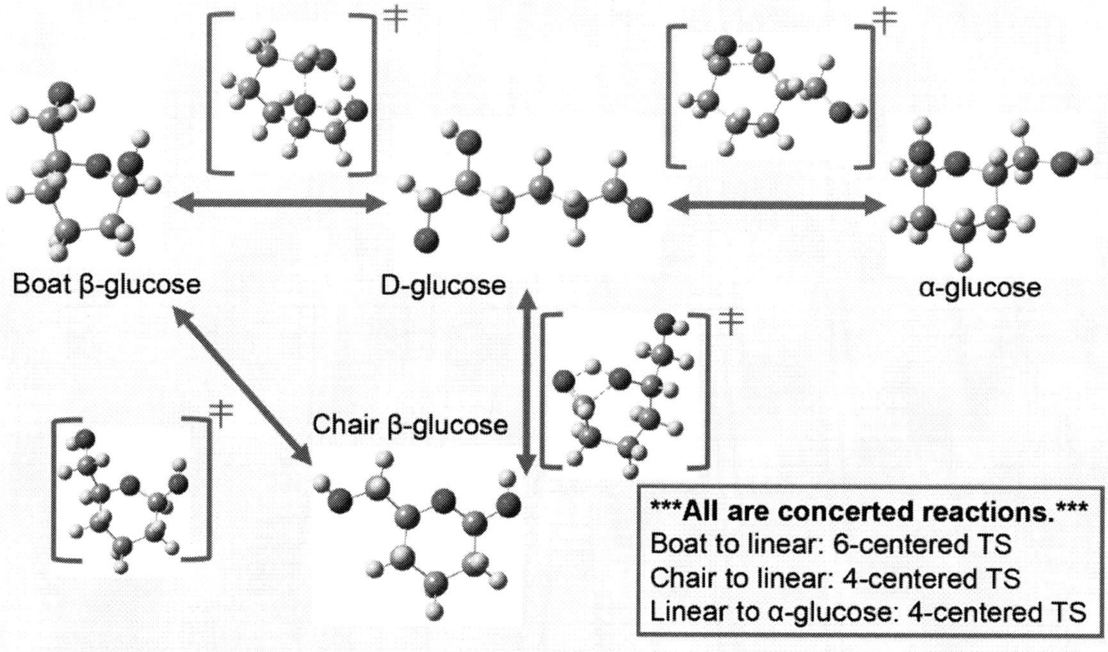

*****All are concerted reactions.*****
Boat to linear: 6-centered TS
Chair to linear: 4-centered TS
Linear to α-glucose: 4-centered TS

NC STATE UNIVERSITY

A bigger surprise: ROH molecule can catalyze!

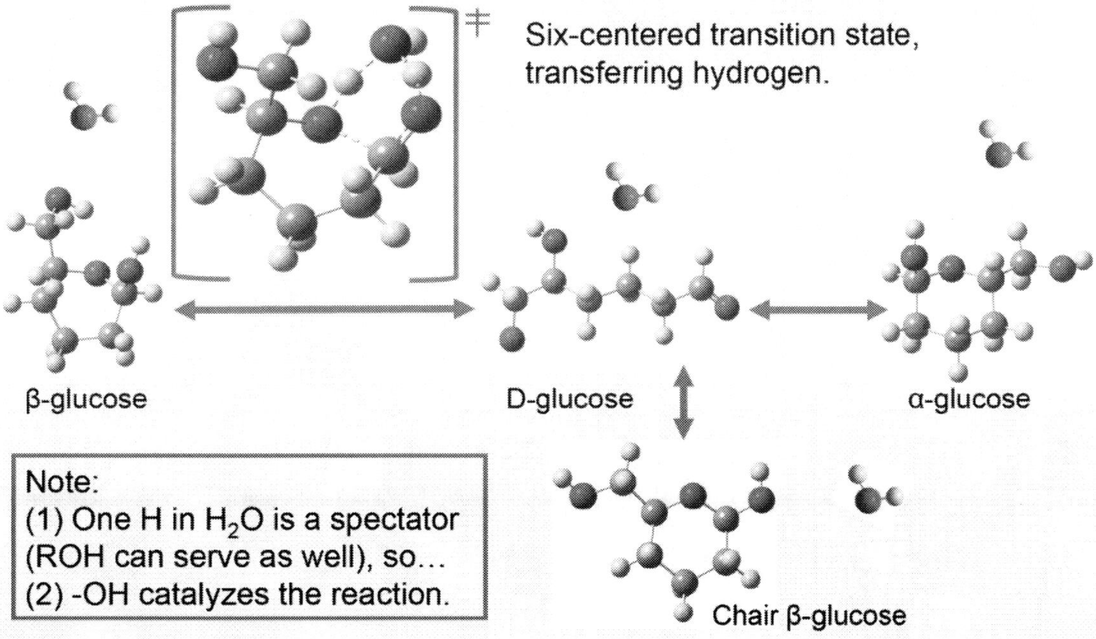

NC STATE UNIVERSITY

Compare the Arrhenius parameters, A in s^{-1} and $s^{-1}atm^{-1}$, E_{act} in kcal/mol.

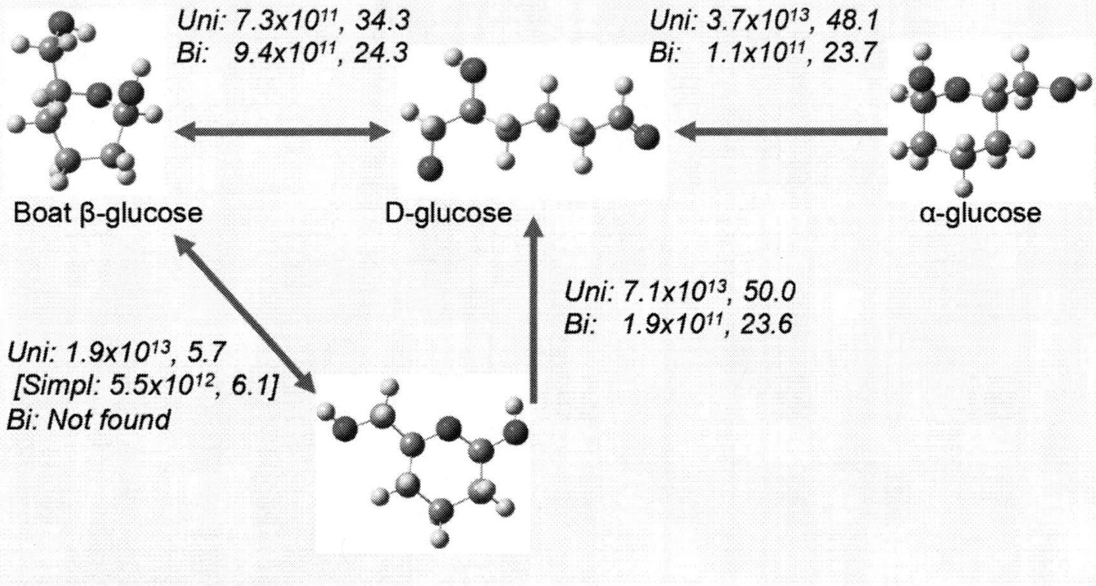

NC STATE UNIVERSITY

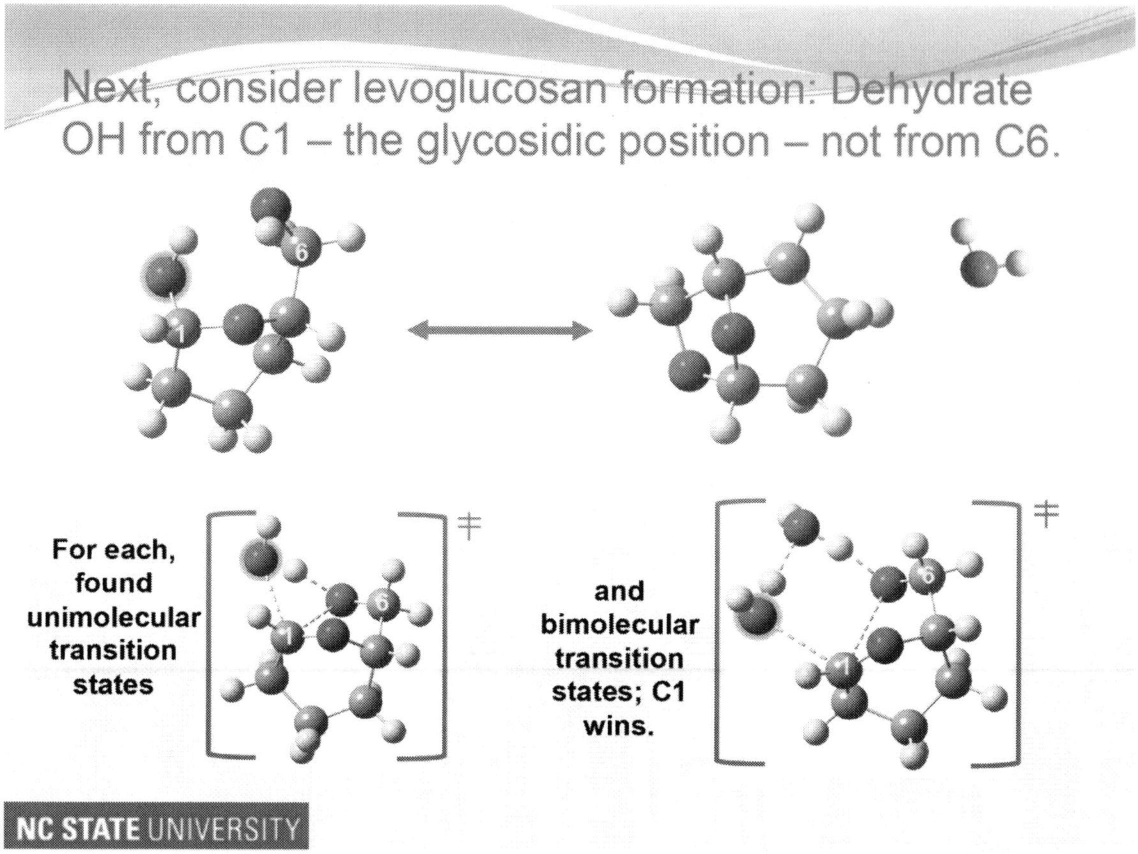
Next, consider levoglucosan formation: Dehydrate OH from C1 – the glycosidic position – not from C6.
6
1
For each, found unimolecular transition states
1
6
‡
and bimolecular transition states; C1 wins.
6
1
‡
NC STATE UNIVERSITY

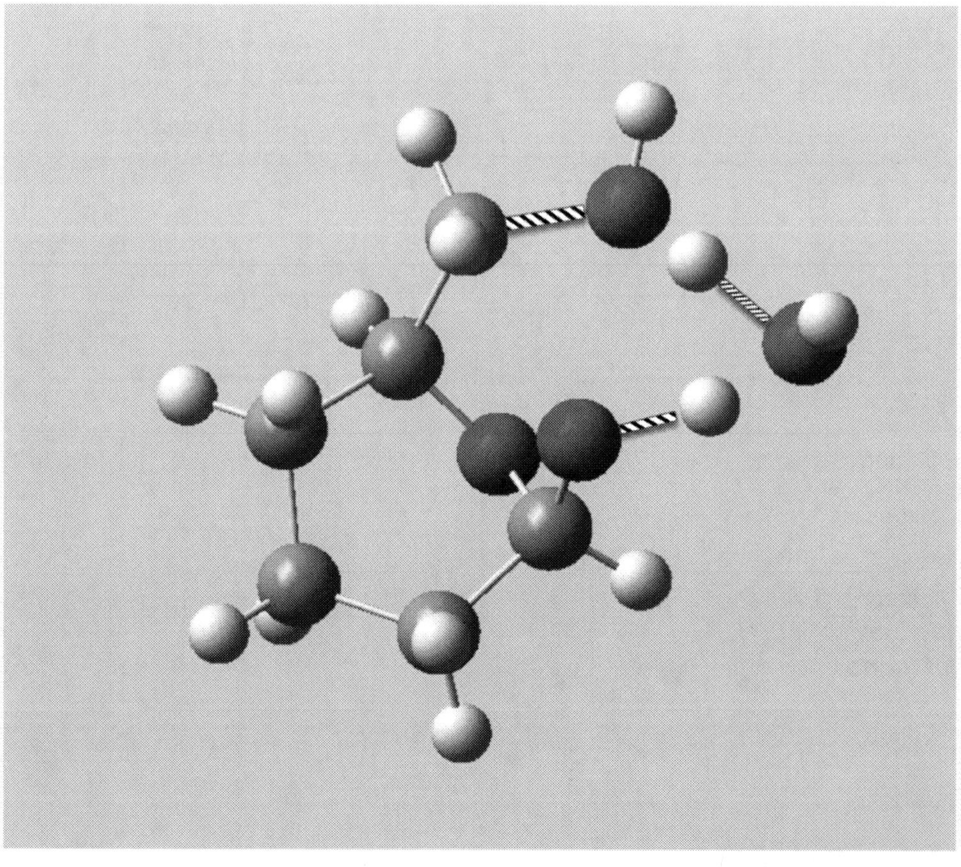

We're moving to hemicellulose. Here, xyloglucan:

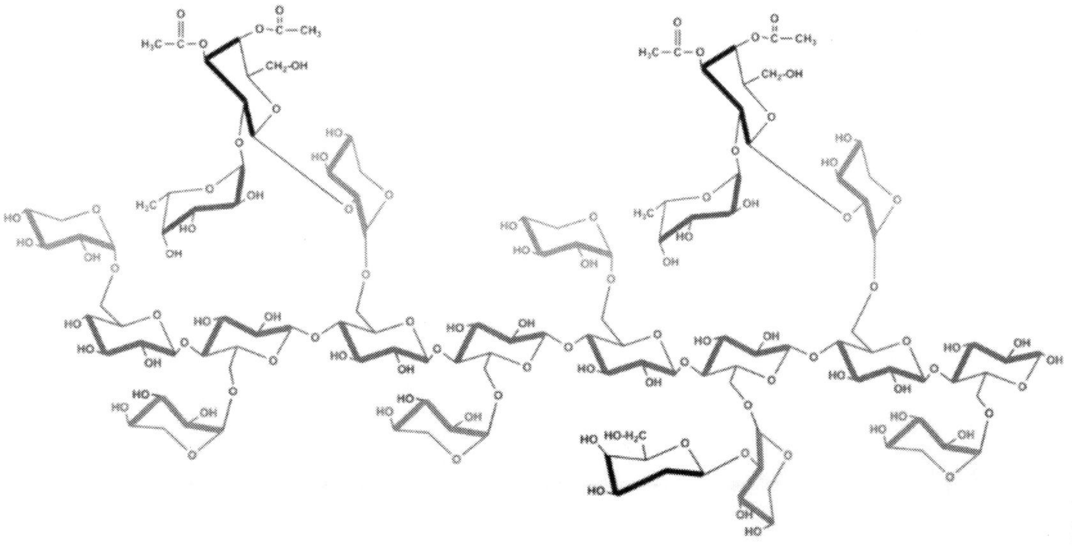

Backbone β-D-glucans; α-D-xylose; **α-D-galactose**; α-L-fucose residues.
M. Ochoa-Villarreal et al. *Plant Cell Wall Polymers*, Ch. 4. DOI: 10.5772/46094

NC STATE UNIVERSITY

From -CH_2OH, levoglucosan can form, and so on.

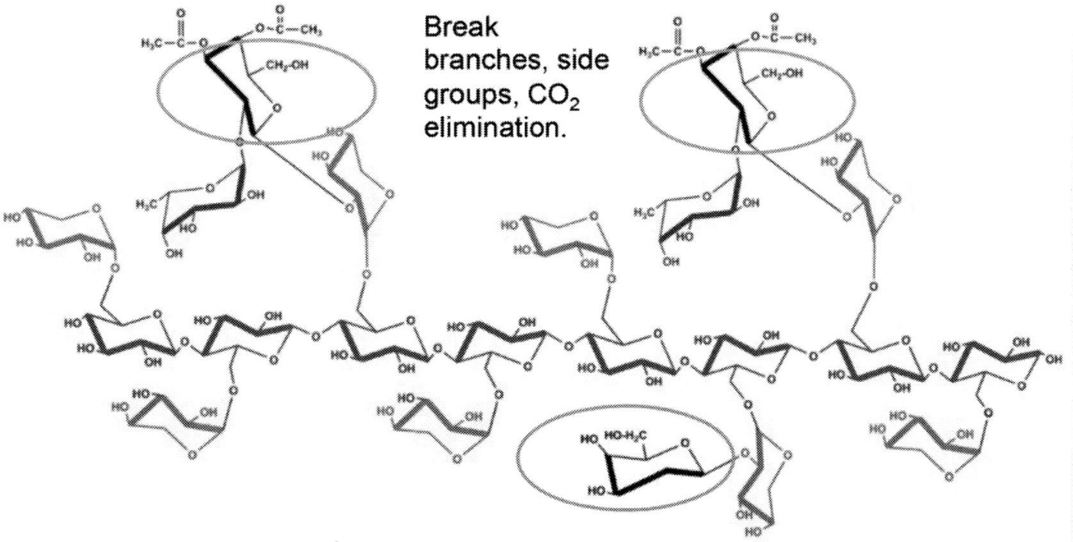

Backbone β-D-glucans; α-D-xylose; **α-D-galactose**; α-L-fucose residues.
M. Ochoa-Villarreal et al. *Plant Cell Wall Polymers*, Ch. 4. DOI: 10.5772/46094

NC STATE UNIVERSITY

Xylose-xylan mimics Glucose-Cellulose.

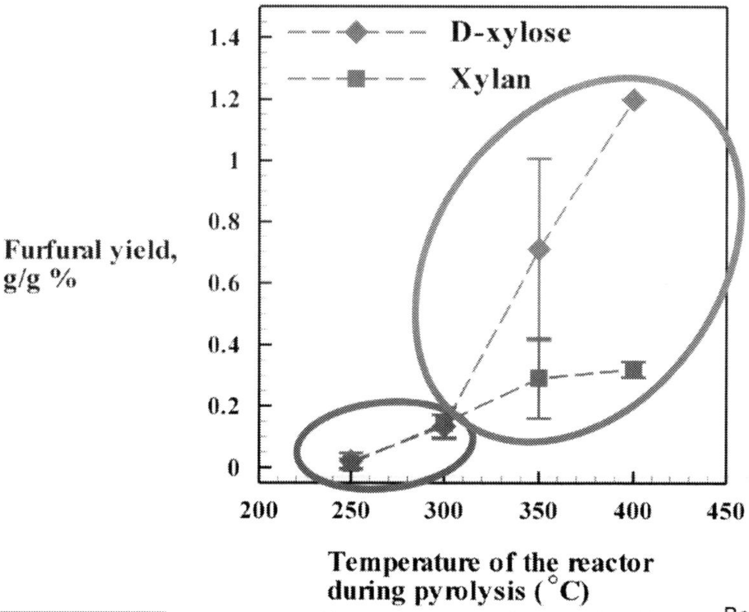

NC STATE UNIVERSITY

Bose and Westmoreland, 2017 AIChE, ppr 446f

The reason is isomerizations.

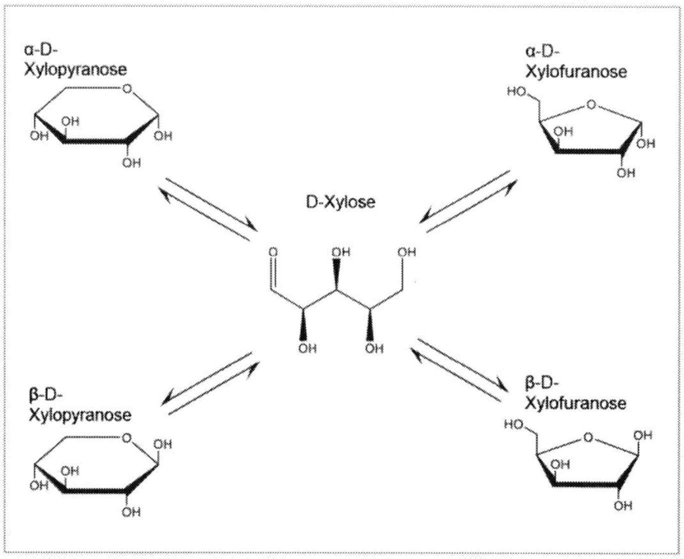

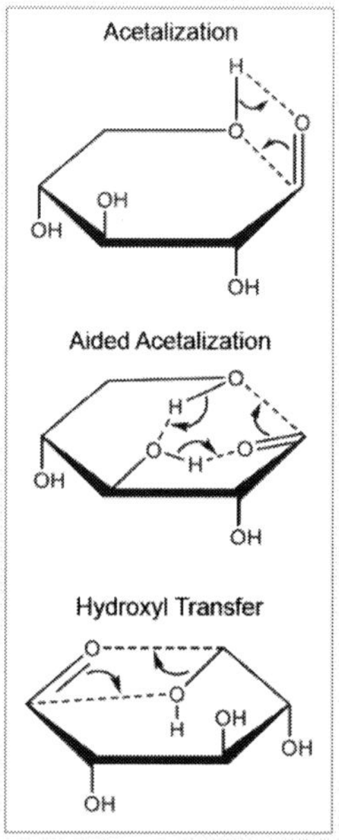

NC STATE UNIVERSITY

McGill and Westmoreland, 2017 AIChE, ppr 174g

So the many possible pathways…

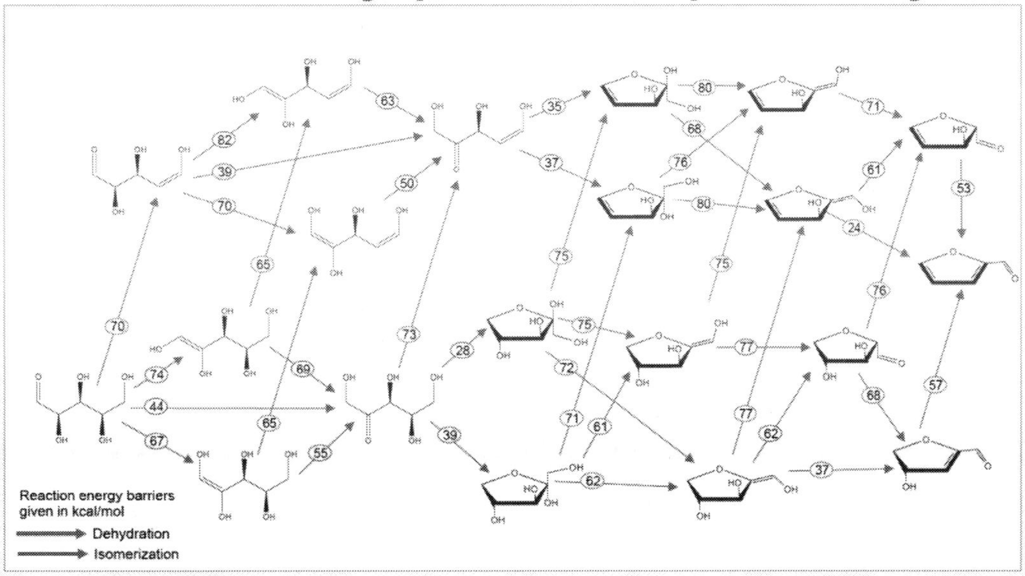

McGill and Westmoreland, 2017 AIChE, ppr 174g

… are dominated by a few.

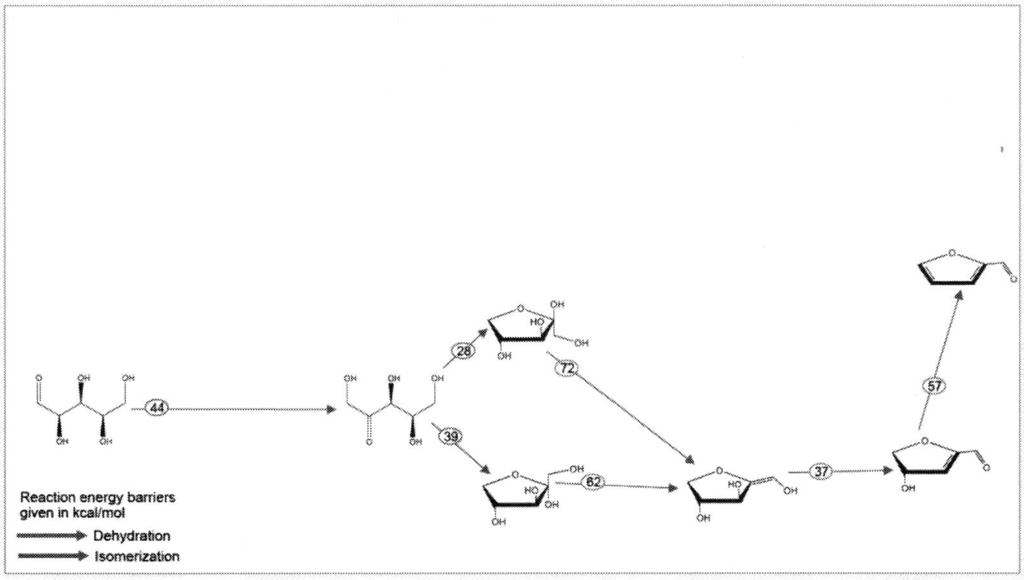

McGill and Westmoreland, 2017 AIChE, ppr 174g

Finally, lignin is "easy".

This is not lignin! Instead, it shows molecular features & linkages.

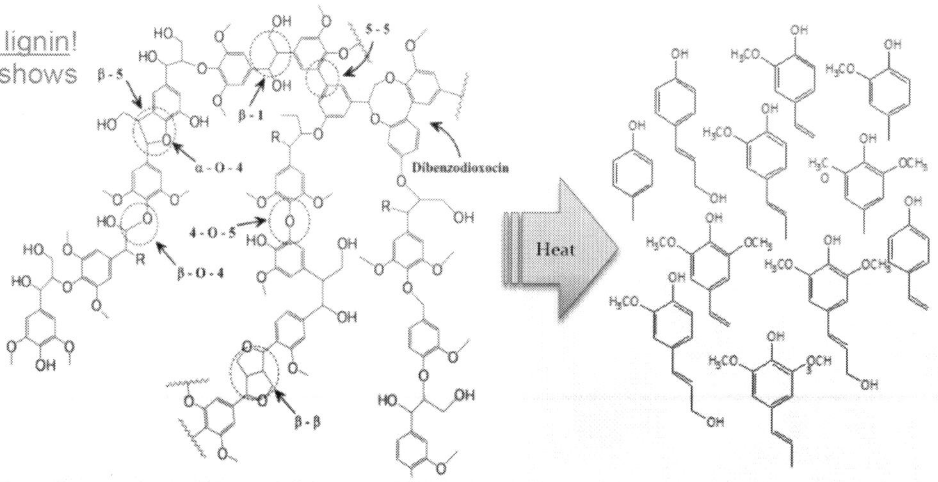

NC STATE UNIVERSITY

Radical routes dominate:

- Experiments and modeling at high temperature:

  M.T. Klein, P.S. Virk, *Ind. Eng. Chem. Fundam.* 22 (1983) 35-45.

  M.T. Klein, P.S. Virk, *Energy & Fuels* 22 (2008) 2175–2182.

  J. Cho, S. Chu, P.J. Dauenhauer, G.W. Huber, *Green Chem.* 14 (2012) 428-439.

  show that pyrolysis is dominated by homolytic bond-breaking to form radicals, which then break down by β-scission and abstractions.

- The chemical steps are amenable to radical-chemistry modeling that has been developed for combustion and thermal cracking of polymers; see:

  T. Faravelli, A. Frasoldati, G. Migliavacca, E. Ranzi, *Biomass and Bioenergy* 34 (2010) 290-301

NC STATE UNIVERSITY

In summary,

- We begin to see – and quantify – many of the species, pathways, and elementary steps occurring in lignocellulosic pyrolysis.
- The likely implementation will be two-fold:
  - Structure-oriented lumping as Quann and Jaffe have done for petroleum.
  - Reactive molecular dynamics (RxnMD from my group, ReaxFF from van Duin) to capture chemistry and physics.

NC STATE UNIVERSITY

Acknowledgements

- We gratefully acknowledge support of this research by the U.S. National Science Foundation / CBET under award 1512401, "Establishing Elementary Mechanisms of Biomass and Hemicellulose Torrefaction."

- **Glucose / cellulose / sugars:** Chas McGill, Arnab Bose, Katie Katie Cao, Vikram Seshadri, Jordan Keith.
- Supercomputers: SDSC comet under XSEDE grant TG-CHE100121 and N.C. State University's Henry2 cluster.

NC STATE UNIVERSITY

485888 Kinetics and Energetics of Biomass Pyrolysis By Pulse-Heated Analysis of Solid Reactions (PHASR)

Abstract:

Thermochemical conversion of lignocellulosic biomass utilizes high temperature chemical decomposition to produce liquids via pyrolysis, solids via torrefaction, or gases via gasification which are subsequently converted to liquid biofuels or biochemical[1]. However, these systems are comprised of thousands of molecules and millions of chemical reactions within three phases, making them intractable by conventional research techniques. In this presentation, utilization of an isothermal biomass kinetics reactor called â€œPHASRâ€ (Pulse-Heated Analysis of Solid Reactions) enables millisecond time-resolved measurement of 100-200 chemical compounds produced from cellulose at 300-550 Â°C[2]. The design of the technique and its performance are presented. The requirements for elimination of transport (heat and mass) artifacts are identified along with experimental measurement identifying the kinetic transition between transport- and reaction-controlled conditions; biomass films reacting at 500 Â°C must be thinner than 70 μm. The experimental kinetics of cellulose conversion indicate a key cellulose initiation mechanism transition at 467 °C, wherein the low temperature intra-chain initiation mechanism ($E_{a,app}$ 23.2±1.9 kcal/mol, k_0 2.0×10^7 s^{-1}) is converted to a high temperature intra-chain initiation initiation mechanism ($E_{a,app}$ 53.7±1.1 kcal/mol, k_0 2.4×10^{16} s^{-1})[3]. The mechanisms of cellulose initiation and further reaction to volatile products are discussed in relation to the measured energies.

[1] **"Top Ten Fundamental Challenges of Biomass Pyrolysis,"** M.S. Mettler, D.G. Vlachos, P.J. Dauenhauer, *Energy & Environmental Science* 2012, 5, 7797-7809.

[2] **"Millisecond Pulsed Films Unify the Mechanisms of Cellulose Fragmentation**," Christoph Krumm, J. Pfaendtner, P.J. Dauenhauer, *Chemistry of Materials* 2016, 28(9), 3108-3114.

[3] **"Energetics of Cellulose and Cyclodextrin Glycosidic Bond Cleavage**," Cheng Zhu, Christoph Krumm, Greg Facas, Matthew Neurock, Paul J. Dauenhauer *Reaction Chemistry and Engineering* 2017. DOI: 10.1039/C6RE00176A

503100 Theoretical Elucidation of the Molecular Behaviours of Key Compounds during Biomass Pyrolysis

Abstract:

Biomass pyrolysis is an efficient way to transform raw biomass or organic waste materials into useable biofuels and/or valuable chemicals. The production process can be described as biomass components decompose at various depolymerization stages based on their stability and operating temperature: firstly hemicellulose, followed by cellulose, and lignin decomposition at the final stage. During this decomposition sequence, thousands of compounds/intermediates (e.g., anhydro-sugars, aromatics, phenols) form, react, remain or disappear in the system. Despite the investigation history on the production of those compounds at the macro scales by both experiments and modellings, the fundamental atomic/molecular level investigations in this area have not been extensively carried out to date.

Levoglucosan (1,6-anhydro-Î²-D-glucopyranose) is representative of one of these compounds. It is one important primary product during cellulose pyrolysis either as an intermediate or as a product. It is initially formed from cellulose decomposition and further converts into low-molecular-weight compounds during pyrolysis secondary reactions. It remains one of the main compounds of raw bio-oil, comprising up to 78 wt%.

Within this talk, the fundamental investigation on the mechanism and kinetic modelling of the production of levoglucosan from lignocellulosic biomass will be presented. Three available mechanisms for levoglucosan formation have been studied theoretically by performing density functional theory based calculations. These mechanisms are free-radical mechanism; glucose intermediate mechanism; and levoglucosan chain-end mechanism. Specifically, the molecular behaviour of levoglucosan has elucidated by revealing 14 reaction pathways, 26 elemental reaction steps, and the involved around 60 compounds act as intermediates, transition states, or products. By comparing with the activation energy obtained from the experimental results, it was concluded that free-radical mechanism has the highest energy barrier during the three levoglucosan formation mechanisms, glucose intermediate mechanism has lower energy barrier than free-radical mechanism, and levoglucosan chain-end mechanism is the most reasonable pathway because of the lowest energy barrier. The variational transition state rate constants for every elementary reaction and every pathway were calculated. The first-order Arrhenius expressions for these elementary reactions and pathways were suggested.

As an intermediate during cellulose pyrolysis, levoglucosan will be easily decomposed as part of secondary reactions of tar. The thermal decomposition mechanism of levoglucosan was also studied using density functional theory methods. The decomposition included direct C-O bond breaking, direct C-C bond breaking, and dehydration. The properties of the reactants, transition states, intermediates, and products for every elementary reaction were obtained. It was concluded that C-O bond breaking is easier than C-C bond breaking due to a lower activation energy and a higher released energy. During the 6 levoglucosan dehydration pathways, one water molecule which composed of a hydrogen atom from C3 and a hydroxyl group from C2 is the preferred pathway due to a lower activation energy and higher product stability.

Except for levoglucosan, the formation pathways of other compounds were also investigated. It was found that formaldehyde can be formed from anhydroglucose which undergoes a hydrogen-donor reaction, the anhydroglucose can be obtained from a cellulose hemolytic reaction during high-temperature steam gasification of the biomass process. At a pressure of 1 atm, levoglucosan can be formed at all of the temperatures (450-750 K) considered in this simulation, whereas formaldehyde can be formed only when the temperature is higher than 475 K. Moreover, the energy barrier of levoglucosan formation is lower than that of formaldehyde. It was also found that 1-pentene-3,4-dione, acetaldehyde, 2,3-dihydroxypropanal, and propanedialdehyde can be formed from the C-O bond breaking decomposition reactions. 1,2-Dihydroxyethene and hydroxyacetic acid vinyl ester can be formed from the C-C bond breaking decomposition reactions.

This research provides thermodynamic assessment of the available routes for the production of key compounds, identifies key challenges and future trends for second-generation bio-chemicals. It also confirms that Quantum Mechanics based simulation can reveal fundamental phenomena, which are difficult to be explored from traditional experimental techniques, and can be used to guide the experimental design and industrial application.

502884 Study of Red Oak Derived Lignin, Pyrolytic Lignin, and Hydrogenated Pyrolytic Lignin with 2D-NMR, Fticr-MS, and GPC

Abstract:

Pyrolytic lignin is a major component of bio-oil derived from the lignin portion of biomass during pyrolysis. Better understanding of the structure of pyrolytic lignin can aid in utilizing this material for fuels or chemicals. Analysis of lignin and lignin-derived streams cannot be done with standard analytical techniques, like gas chromatography, because it is non-volatile. In this talk, we will discuss the results from the analysis of the structure of red-oak lignin before and after pyrolysis and hydrogenation with FTICR-MS, GPC, and NMR. The lignin was isolated from red oak with a biomass fractionation method utilizing gamma-valerolactone as a solvent. Pyrolysis was performed in a fluidized bed pyrolysis reactor operated at 500 °C and preheated nitrogen flow. An attached bio-oil collection system separated the bio-oil into five stage fractions. The pyrolytic lignin was separated with a water fractionation from the second stage fraction. Hydrogenation was performed in an ethanol solution over a 5 wt% Ru/C catalyst at 150 Â°C and 500 psig of initial H_2 for 2.5 h. A five times decrease in the average molecular weight (from 4660 Da to 831 Da) was observed with GPC after pyrolysis, further decreasing to 726 Da after hydrogenation. More than 1100 distinct molecular weights were observed with FTICR-MS of the lignin stream. The FTICR-MS revealed O/C and H/C ratios decreased from 0.33 and 0.94 to 0.25 and 0.90 after pyrolysis, in line with dehydration of sidechain â€"OH to alkenes while the O/C ratio decreased to 0.23 and the H/C ratio increased to 1.11 after hydrogenation, supporting the hydrodeoxygenation and saturation of the aromatic units. Quantitative 13C NMR revealed a large decrease in the C-O aliphatic carbons (from 35.0% to 11.4% of carbons) during pyrolysis with a large increase in carbonyl and C-C aliphatic carbons (from 0% and 1.3% to 4.2% and 15.4% of carbons respectively). The hydrogenation decreased the carbonyl and aromatic carbons (from 4.2% and 69.0% to 2.6% and 37.0%) and increased the C-C and C-O aliphatic carbons (from 15.4% and 11.4% to 41.3% and 19.2% respectively). The methoxyl region decreased from 13.3% to 6.3% of carbons after pyrolysis and 4.2% of carbons after hydrogenation. HSQC NMR displayed the varying amounts of C-C aliphatic, C-O aliphatic, aromatic, and aldehyde functionalities. From pyrolysis, evidence supported the formation of diaryl methine linkages while the lignin Î²-O-4, phenylcoumaran, and resinol linkages disappeared. Hydrogenation produced cyclohexyl functionalities through saturation of the aromatics. Benzaldehyde and cinnamaldehyde were seen for the lignin and pyrolytic lignin but completely disappear after hydrogenation. These aldehyde functionalities were highly reactive during hydrogenation and potentially contributed to coke formation through aldol condensation reactions. With HMBC NMR, the lignin had small amounts of C-C aliphatic to aromatic correlations while these correlations increased after both pyrolysis and hydrogenation. C-C to C-O aliphatic and C-C to C-C aliphatic correlations increased after hydrogenation. Tertiary aromatics and C-C aliphatic to carboxylic acid correlations were present in lignin, pyrolytic lignin, and hydrogenated pyrolytic lignin. Ketone to C-C aliphatic correlations appear after pyrolysis but similarly to the aldehydes, disappear after the hydrogenation. The results from this presentation provide a more comprehensive understanding of the chemistry that occurs during lignin pyrolysis and hydrogenation.

503169 Overview of Issues Around Deconstruction of Lignin

Abstract:

There is widespread interest in the development of large-scale, economically competitive biorefineries that combine integrated catalytic, thermal, and biological conversion processes to efficiently utilize all of the materials and energy contained in biomass. In commercial large-scale biorefineries, stringent requirements for process performance and profit/cost metrics will have to be met. However, utilizing biomass-derived carbohydrates for bioethanol production generates similar amounts of lignin as a by-product which is either discarded or burned to recover low-grade heat. To improve the economic competitiveness and reduce the environmental footprint of such biorefineries, technologies are needed that utilize both carbohydrate and lignin fractions of biomass and that ultimately improve biomass-based product yields and selectivities. Simply stated, this implies obtaining greater value from lignin. Successful research and development activities for catalytic lignin deconstruction will require three-dimensions of integrated understanding: (1) catalyst design and characterization, (2) reaction and process engineering, and (3) lignin biosynthesis, structure, extraction, chemistry, and characterization. Often in lignin catalysis, well-designed catalysts are effective on certain model compounds that represent lignin dimers (usually Î²-O-4 dimer), but fail to be effective on real lignin To solve such issues, research must rely on a deep expertise in (1) characterizing biomass and lignin, (2) building accurate (more complex dimers, oligomers, and polymer systems) lignin models, (3) applying methodologies to resolve the complex mixtures that tend to result from lignin deconstruction, and (4) conducting operando characterization of high temperature and pressure reaction condition. I will present an overview on the issues concerning the deconstruction of lignin with a particular focus on reductive methods for the selective disassembly of lignin with heterogeneous catalysts and hydrogen.

502340 Mesoscale Modeling of Thermochemical Conversion of Biomass

Abstract:

The outcomes of thermochemical conversion processes are governed by an intricate coupling of chemical reaction kinetics and complex transport phenomena. Computational tools provide an excellent complement to experiment for the purposes of process optimization and interpretation of experimental results. While much attention has been devoted to modeling conversion processes at extreme length scales (i.e., the molecular scale and the bulk process scale), relatively little effort has been focused on understanding the role of lignocellulose mesostructure via computational modeling. In the context of this presentation, the mesoscale architectural features of biomass are defined as the higher-order assembly of biopolymers in the plant cell wall and the species specific microstructure present in biomass particles. In this presentation, I will present general methodology by which the impact of structural features of lignocellulose may be assessed in silico. A general method based on constructive solid geometry will be presented that can be used to explicitly model lignocellulosic biopolymer assemblies on the length scale of ~ 100 nm. Several examples of how quantitative microscopy and chemical analysis can be used to parameterize the model for various scenarios will be included. Next, I will describe how morphologically accurate 3D biomass particles may be constructed and employed in finite element simulations of heat and mass transport coupled to conversion kinetics. Finally, I will present an example of how this methodology may be employed to estimate optimal conversion conditions for specific biomass feedstocks.

503580 Fire-Adapted Microbes

Abstract:

Efforts to convert plant biomass into fuels and other bio-based products are increasingly focusing on thermal methods. These methods offer several key advantages over enzymatic deconstruction routes, including being 'agnostic' to the type of biomass being treated. Heat-modified residues, however, still require enzymatic processing, and current enzyme 'cocktails' are not optimized to these tasks. These residues may also contain lignin residues that repolymerize after initial depolymerization, a chemical process that in nature can be overcome biochemically via oxidative enzymes such as peroxidases. Our goal is to identify and validate novel biomanufacturing enzymes from a logical organism source - fire-adapted microbes.

Fungi are perhaps the most well-studied examples of fire-adapted microbes, and an outstanding example is *Neurospora crassa*. This Ascomycete fungus has been co-opted by those studying the gene-for-enzyme theory as a model system for genetic recombination, using its easily-counted melanized spores. This fungus has also been a key model system for studying circadian rhythms, tracking daily sporulation 'rings' not controlled by light. *Neurospora* has become an enzyme toolkit for studying its habits out of the context of its natural niche on fire-killed plant biomass. This essentially leaves a treasure trove of genetics tools disconnected to a key trait of the fungus, and the gains in unlocking this system in context of thermal deconstruction routes look very promising. There are also other fungal genera (e.g. *Gloeophyllum*, *Schizophyllum*, *Antrodia*) associated with fire, many of which have newly-annotated genomes in the Joint Genome Institute portal, and elsewhere.

Tapping into these fire-adapted microbial systems may reveal genes with more relevance to thermal deconstruction processes and with more potential for tailoring to biomanufacturing processes 'downstream.' Specifically, these pathways may be more efficient than current cocktails in their hydrolysis of anhydro oligomers, detoxification of thermally-derived fermentation inhibitors (e.g. acetate), and stabilization of solubilized lignin. Traditional cocktail enzymes were not designed for this job, and their catalytic domains did not evolve on thermally-modified substrates such as levoglucosan. Fire-adapted fungi are a logical prospect to address these needs. Given the range of possibilities for fire-adapted microbes, my talk will present the range and traits of these uniquely-adapted organisms, and the genomic toolkits for tapping into their biotechnologically-relevance capacities going forward.

503773 Rewiring Yeast to Improve Cellobiose Fermentation

Abstract:

Sugars derived from plant cell walls are highly abundant in nature and hold promise to provide an abundance source for production of renewable fuels and chemicals. However, the cost and logistics of releasing these sugars from plant biomass remains prohibitive on commercial scales. The yeast *Saccharomyces cerevisiae* has a proven track record as a research platform and industrial host for production of biofuels and chemicals. Yet, *S. cerevisiae* is unable to utilize sugars released from the plant cell wall, i.e. cellobiose. Using pathways from *Neurospora crassa*, we have engineered *S. cerevisiae* to ferment cellobiose. However, having never used cellobiose as a carbon source, *S. cerevisiae* ferments this "masked" source of glucose slowly. We have used *S. cerevisiae* as a "blank slate" to understand the key bottlenecks in cellobiose utilization. I will present our recent work that combines directed evolution, pathway engineering, and systems biological methods to further improve the fermentation performance of cellobiose-utilizing *S. cerevisiae*.

497574 Engineering Strategies for Improving Microbial Utilization of Thermally Depolymerized Biomass

Abstract:

Thermochemical processing is an attractive means of depolymerizing biomass into sugar and aromatic monomers for biological upgrading to fuels and chemicals. However, microbial utilization of these compounds is limited at several levels. Most of the commonly-used microbial biocatalysts lack the metabolic pathways needed for utilization of the most abundant anhydrosugars, levoglucosan and cellobiosan. Even though these anhydrosugars comprise a substantial fraction of the global carbon cycle, microbial utilization of these sugars has remain largely undercharacterized until recently. We will discuss efforts to characterize these pathways and to engineer production organisms for utilization of these sugars. A second challenge is the presence of trace inhibitory compounds in the depolymerized biomass. This problem can be addressed by chemical treatment of the sugars to remove the inhibitors, or by engineering of the microbial biocatalyst for improved tolerance â€" both approaches will be described. Finally, we will discuss efforts to improve microbial utilization of aromatic monomers produced from the lignin fraction.

486640 Production of Fermentation Substrates from Thermal Deconstruction of Lignocellulose

Abstract:

Fast pyrolysis of biomass utilizing a condensation train that condenses and collects pyrolysis products according to their vapor pressures has permitted the separation and purification of multiple bio-oil streams. Heavy ends bio-oil derived during fast pyrolysis of lignocellulosic and carbohydrate fractions of biomass can be separated by liquid-liquid extraction with water for the effective recovery of sugars (water-soluble) and phenolic oil (water-insoluble). While the light ends bio-oil contains the vast majority of water, we have the ability to separate the water from the organics. Our research has shown that the separate recovery of these streams and removal of toxic inhibitors allows for effective fermentation substrates. These substrates are obtained from both heavy and light ends. The sugar stream recovered from the heavy ends can be cleaned using polymeric resin absorbents, SP207 or XAD-4. These resins adsorb the phenolics remaining in the water-soluble sugar stream, ultimately producing a clean stream of sugars. This stream can be hydrolyzed and used in fermentation to butanol or ethanol or directly fermented with ethanol-producing *Escherichia coli*. The water-insoluble fraction from the heavy ends can be further separated into monomeric phenols and larger molecular weight phenols with liquid-liquid extraction utilizing toluene as the solvent. Exploitation of nontoxic emulsifiers has permitted utilizing the monomeric phenol fraction, obtained using toluene, for potential biochemical transformation, by *Pseudomonas putida* KT2440, to muconic acid. Light ends contain carboxylic acids, water and other light oxygenates. The carboxylic acids produced during pyrolysis can be separated from the water stream using ionic resin or membrane technology, both hydrophobic and hydrophilic, producing a clean stream of water that can be easily be disposed of. The acetic acid, by far the highest concentration of carboxylic acids in light ends, is then used in fermentation to methane for process heat.

484665 Biological Conversion of Thermochemical Wastewater Streams

Abstract:

Thermochemical biomass conversion processes such as fast pyrolysis and catalytic fast pyrolysis produce aqueous waste streams that contain aromatic, acid, and aldehyde compounds. These aqueous waste streams often comprise a significant fraction of biomass-derived carbon, and are currently slated for wastewater treatment, representing a cost burden to the thermochemical biorefinery. Prior to developing a microbial valorization strategy, we first conducted a comprehensive analytical survey of thermochemical aqueous waste streams from fast pyrolysis and catalytic fast pyrolysis, identifying and quantifying over 200 compounds at near-complete mass closure. This detailed analytical effort enabled the ability to tailor a microbial strategy for conversion of the wide range of aromatic, acidic, and aldehyde compounds to value-added chemicals using an engineered, aerobic monoculture strategy. Specifically, we are engineering a robust aromatic-catabolic microbe, *Pseudomonas putida* KT2440, to upgrade aqueous streams from pilot-scale pyrolysis operations. Beyond a substantial expansion in substrate catabolism, we have also demonstrated that over-expression of the native protein quality control machinery enables a dramatic improvement in microbial tolerance, a key challenge for the toxic streams produced from thermochemical processing of biomass.

499712 Thermal Deconstruction Opens Biomass for Acid Hydrolysis to Monomeric Sugars

Abstract:

Thermal deconstruction involves rapidly heating then cooling biomass in an inert environment before complete pyrolysis can occur. While some bio-oil and light gases are produced, solid or liquid phase reactions greatly modify the biomass. As shown by microscopy and chemical analysis, thermal deconstruction cracks the biopolymers and loosens their lignocellulosic structure. This change allows molecules to diffuse more readily into and out of the biomass. Furthermore, as the biopolymers fragment into smaller oligomers they become more susceptible to biological decomposition. Three routes for processing both thermally deconstructed cellulose (as a model compound) and biomass were investigated: enzymatic hydrolysis, direct yeast fermentation, and anaerobic digestion.

503323 Ionic Liquid Pretreatment of Lignocellulosic Biomass: Effect of Biomass Composition on Pretreatment Efficiency

Abstract:

Ionic liquid pretreatment of lignocellulosic biomass: Effect of biomass composition on pretreatment efficiency

Globally, bio-fuel is becoming an attractive substitute for fossil fuels due to the increasing demand for clean energy and rapidly depleting crude oil reserves. Utilization of non-food biomass has been recently has been preferred because of its non-competitiveness towards food crops and its generation in the huge amount as a waste by-product. For production of biofuel, lignocellulosic biomass waste must be pretreated in order to break down its complex structure. Pretreatment with ionic liquid (IL) 1-ethyl-3-methylimidazolium acetate ([EMIM][OAc]) has been reported to disrupt structure and dissolve biomass. In this study, the performance of [EMIM][OAc] was investigated for dissolution of four different lignocellulosic biomasses (figure 1). The feedstocks chosen in this study were spent coffee grounds (SCGs), coffee husks (CH), *Eucalyptus globulus* (EG) bark, and cotton gin (CG), which are widely produced as food and industrial waste by-products. These feedstocks significantly differ in their relative composition of cellulose, hemicellulose and lignin. The pretreated biomass was enzymatically hydrolysed by commercial cellulase enzymes and the sugars thus obtained were fermented using wild-type yeast. A comparative analysis of the impact of IL treatment on the feedstocks was carried out, which showed that there was an increase in the digestibility of the pretreated biomass. Highest digestibility was observed in cotton gin, which was the feedstock with the least amount of lignin. The recovery and recycle of the IL was also investigated in this study.

503357 Enabling Microbial Utilization of Lignin-Derived Monomers

Abstract:

Biomass contains lignin, a polymer of phenolic molecules, which provides stability and protection to the plant. Many economic models indicate that the addition of value to lignin is essential for the economic viability of the conversion of biomass to renewable fuels and chemicals. Some microorganisms, such as *Pseudomonas putida*, can convert or metabolize some aromatic and phenolic molecules. Previous reports have shown that *P. putida* can be engineered to funnel multiple phenolics through its central metabolic pathways [1]. A funneling approach can be advantageous because thermally decomposed lignin can be composed of hundreds of different molecules many of which can be aromatic or phenolic in nature. We have identified the following phenolics in our thermally decomposed lignin: phenol, guaiacol, 2-ethyl phenol, 4-ethylguaiacol, 4-vinylphenol, 4-vinylguaiacol, syringol, and vanillin. Phenolic-rich fractions of pyrolyzed biomass have low solubility and therefore cannot be easily accessed by microorganisms in aqueous cultures. In addition, microorganisms can be negatively affected by inhibitors present in processed biomass streams. Here, we have developed emulsions of phenolic monomer-rich product using surfactants or sonication. *Pseudomonas putida* KT2440 grew in emulsions produced using surfactants and phenolic monomer-rich product. Studies suggest that the emulsion allowed *P. putida* to access molecules with low solubility in aqueous culture. In addition, the emulsion produced using surfactants seemed to have an additional benefit of decreasing the toxic effects of the phenolic molecule *p*-coumarate. *P. putida* also grew in emulsions of the phenolic monomer-rich product produced by sonication. The total carbon decreased over time indicating that *P. putida* was metabolizing components of the phenolic monomer-rich product. Ongoing work involves tuning the pyrolysis products to match the products preferred by *P. putida*. In addition, *P. putida* will also be engineered to better funnel the products present in the pyrolysis stream.

Refs: 1. Johnson, C. W.; Beckham, G. T. *Metabolic Engineering* **2015,** 28, 240-247.

503370 Comparison of Product Distribution, Content and Fermentability of Biomass in a Hybrid Thermochemical/Biological Processing Platform

Abstract:

Thermochemical processing is a promising method for the rapid depolymerization of biomass into fermentable sugar and lignin monomers. This study investigated different types biomass, including switchgrass, corn stover, red oak, hybrid poplar, and loblolly pine. The contents of cellulose, hemicelluloses, lignin, nitrogen, phosphors, sulphur, as well as metal elements in these biomasses were determined. The yields of pyrolytic bio-oil, bio-char, and bio-gas from these biomasses were compared. The profile of sugar and contaminants in the obtained pyrolytic sugar solution was analyzed, and a variety of contaminating compounds were identified and quantified, but many others were not identified. The fermentability of pyrolytic sugars derived from these biomasses were investigated. It is also revealed that furans are the largest portion of contaminants, and their concentrations in the pyrolytic sugar are more than 50% of the total identified contaminants, for all the investigated biomasses. The second largest portions of contaminants are aromatic compounds. This work provides insight into the selection of biomass types for biorefinery models and other analyses.

503383 Low Temperature Hydrogenation of Pyrolytic Lignin over Ru/TiO_2 : 2D HSQC and ^{13}C NMR Study of Reactants and Products

Abstract:

Pyrolysis is a low cost method for converting lignocellulosic biomass into a liquid feedstock, in this case bio-oil. Due to high oxygen content and unsaturated compounds, hydrotreating of bio-oil is essential to produce a usable fuel, though it suffers from high amounts of coke formation causing catalyst deactivation and reactor plugging. Pyrolytic lignin is thought to be a culprit in the coke formation while little is known about the compounds responsible or how to prevent it. Low temperature hydrogenation (25-150 °C) of a mixed maple wood pyrolytic lignin over Ru/TiO_2 catalyst was studied to determine the amount of coke formation on the catalyst surface. HSQC and quantitative ^{13}C NMR was used to elucidate changes in carbon functionalities of the products. Coke formed on the catalyst surface (1% carbon yield to coke) even for hydrogenation at 25 °C. The carbon yield to coke increased from 1% to 5% as the hydrogenation temperature increased from 25 to 150 °C. A single-step hydrogenation at 150 °C decreased the aromatic carbon from 65% to 39% while a three-step hydrogenation scheme at 150 °C decreased the aromatic carbon from 65% to 17%. The decrease in the aromatic carbon corresponded with an increase in the aliphatic carbon. Coke formation reduced from a 4.9% carbon yield in the first hydrogenation step to a 1% carbon yield in each of the second and third hydrogenation steps. The pyrolytic lignin was phase separated into a low and high molecular weight fraction and separately hydrogenated at 150 °C. The low and high molecular weight fractions had a 3.9% and 7.7% carbon yields to coke, respectively.

503019 One-Step Fermentable Sugar Production from Lignocellulosic Biomass Via Soluble Magnesium Salt-Based Catalyst

Abstract:

One of the main obstacles for the effective utilization of lignocellulosic biomass is the considerably high cost for their deconstruction and hydrolysis to release simple soluble sugars. Cost-effectively converting biomass to sugar not only could reduce overall cost of the biomass-to-fuel production; but also might facilitate subsequent conversions of biomass to other biobased products. Both catalytic and enzymatic methods have been developed for biomass-to-sugar conversions, but many of them require additional steps. Here, in this study, a novel strategy for one-step lignocellulosic biomass-to-sugar conversion via soluble salt-based catalyst system was reported. Based on investigating the salt-base materials and their catalytic effects on C-C and C-O bond cleavages, we identified a class of soluble magnesium salt-based catalysts to facilitate biomass deconstruction and hydrolysis technology by directly conversing hemicellulose and cellulose in lignocellulosic biomass to soluble fermentable sugar under the mild acidic condition (pH = 4.3 ~ 5.0) at temperature range 180 ~ 250 ^{0}C. In comparison to conventional concentrated or dilute sulfuric acid process, our technologies have several superior advantages: (1) It is one step to break down lignocellulosic biomasses and to catalytically form sugars, so it does not require enzyme for hydrolysis and it is therefore characterized as a single step mild-acid catalytic process; (2) We can easily control catalyst acidities and phase behaviors by varying catalyst compositions and reaction temperatures. This could facilitate product separations and reduce wastewater environmental concerns; (3) Our studies of ethanol fermentation by *Saccharomyces cerevisiae* shown that our catalysts did not have significant effect on cell growth and ethanol production, therefore, they have great potentials to be integrated with downstream application platforms such as bioconversion of sugar to ethanol or other biobased products. (4) The main catalyst composition of our catalysts, $MgCl_2$, is extremely cheap (less than $100 USD/ton), which is essential for the development of the cost-effective processes. In view of these advantages, we believe our developing catalyst systems possess great technical values for effective biomass utilization and demand in-depth studies for their commercialization feasibilities.

498227 Selective Conversion of Ashe Juniper Waste into Levoglucosenone and Acetol

Abstract:

Ashe juniper is a problematic plant in central Texas due to rapid expansion causing the alteration of rangeland plant communities and hydrological landscapes. Current Ashe juniper waste management is less economical. Thermochemical conversion of Ashe juniper waste into high-value chemical is one way to valorize the waste. This study presents the effects of metal cations and lignin contents on the yield and selectivity of levoglucosenone (LGO) and acetol (AC) by pyrolysis of Ashe juniper waste. Kinetic model was developed to evaluate faster reaction pathway on each condition. It was confirmed that the yield of LGO and AC was highly dependednt on the type and concentration of metal cation and lignin contents. Temperature was the other crucial factor to control the pathway to desirable products relative to undesirable products. Finally, solvent extraction method was tested to maximize the yield of LGO and AC.

IOWA STATE UNIVERSITY

Bioeconomy Institute

AIChE Annual Meeting 2017
Topical Conference: Thermal Deconstruction of Biomass
November 1st, 2017

Arpa Ghosh, Wenqin Li, Mark Mba Wright, Robert C. Brown

Poster: 639g

Fermentable Sugars from Woody Biomass Using THF/Water and Acid Catalyst

Introduction

Challenges of conventional enzymatic pathway

- Cost of enzyme production
  - Covers 20% of total production cost of fermentable carbohydrate
  - Estimated cost contribution up to $1.47 per gallon of ethanol
- Slow rates of bioconversion
  - It takes days or even week to reach glucose yields > 70%
- Attaining high concentration fermentation titer
- Substrate specific enzyme production may be uneconomical

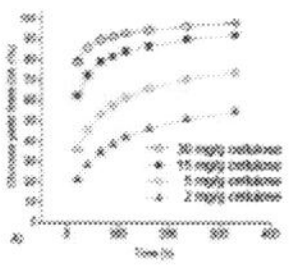

DA: Dilute acid pretreatment (common in biochemical processing of cellulosic biomass, Nguyen, 2015)

Non-enzymatic route with polar aprotic solvents

- Accelerated rate of decomposition of biomass to carbohydrate monomer in polar aprotic solvents
- Dilution prevents secondary reactions
- Mild temperature (150-350 °C)
- Carbohydrates are solubilized
- Low acid consumption

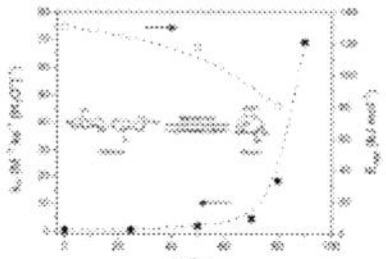

Enhancement of monosaccharide production rate in presence of polar aprotic solvent GVL (Mellmer, 2014)

Recent progress in field:

- Production of 68% theoretical soluble carbohydrates yield from maple wood using gamma-valerolactone (GVL) and water with acid catalyst (Luterbacher et al., 2014)

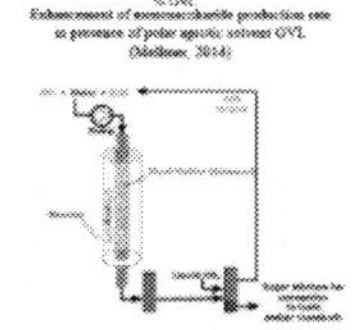

Overcoming Solvent Liquefaction Barriers

Barriers in GVL-mediated process

- Commercially available GVL is expensive and thus demand quantitative recovery
- Boiling point of GVL is high (207-208 °C) making it highly difficult to separate from sugar solution by simple distillation
- Recovery process of GVL is complicated and costly due to the use of a series of liquid CO_2 extraction units

Approach using THF/Water with dilute sulfuric acid

Step 1. Solubilize lignin and hemicellulose by mild pretreatment
Step 2. Deconstruct cellulose-rich pulp to soluble sugars by acid-catalyzed THF-Water co-solvent

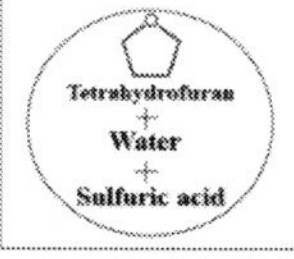

Why THF?

- Easy to separate by flashing (b.p. is 66 °C)
- Lower cost ($121/kg of THF, $524/kg of GVL)
- Good solvent for extracting lignin as soluble
- Depolymerizes cellulose effectively at mild conditions with water and dilute acid catalyst

Process Development

Two-step solvent liquefaction of red oak using THF/Water

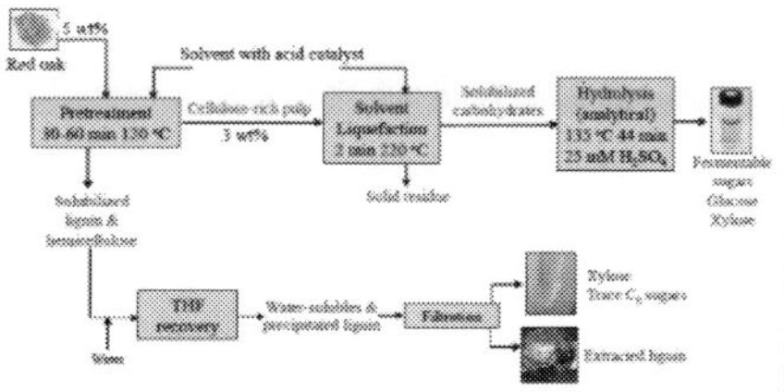

Solvent: THF/Water 80/20 vol%
Catalyst: 2.5-10 mM H_2SO_4 (0.5-2 wt% of red oak) for pretreatment, 2.5 mM for solvent liquefaction

Fermentable Sugar Production

- Over 44 wt% of biomass converted to total solubilized carbohydrates (72% of theoretical yield)
- Pretreated red oak showed high yields of solubilized carbohydrates mainly attributed to dramatically increased glucose yields from pretreated biomass
- Glucose and xylose were obtained at 67% and 96% theoretical yield from two-step solvent liquefaction process
- Biomass was solubilized at 68 wt% with polysaccharide conversion of 97% of theoretical
- Dehydration of monosaccharide resulted in small amounts of furfural and 5-HMF (4 and 5 wt% on biomass basis, respectively)

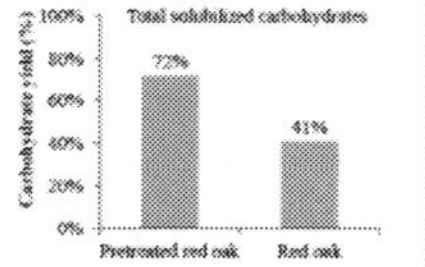

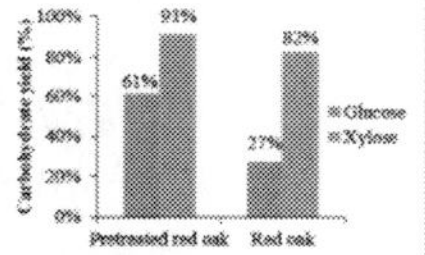

Impact of Delignification

Delignification enriches cellulose in pretreated red oak

- Removal of lignin and xylan from enriches red oak with glucan
- Cellulose-rich red oak is more susceptible to deconstruction by solvent and acid

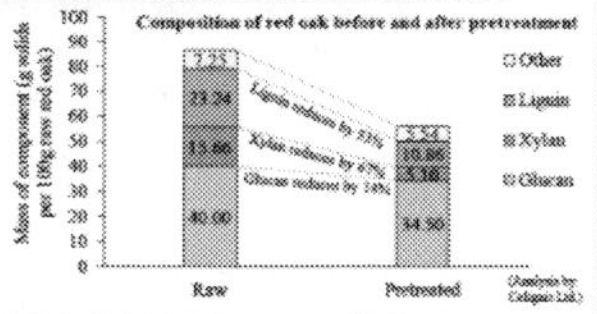

Delignification correlates to sugar recovery

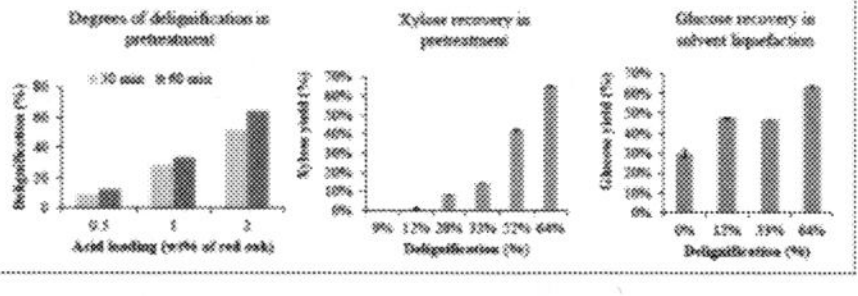

Structure of delignified biomass

- SEM images show polymeric matrix of biomass was loosened up by pretreatment
- More penetrable structure with large pores and microgrooves
- Wrinkles around edges imply loss of structural strength; attributed to delignification

Untreated red oak   Pretreated red oak

Solvent Separation and Sugar Recovery

- Solubilized carbohydrates were recovered at high concentration in water fraction of solvent (80/20 vol% THF/Water) by flashing THF
- Fermentable sugar solution from distillation was rich in glucose, xylose and levoglucosan but had no furans or phenolic compounds

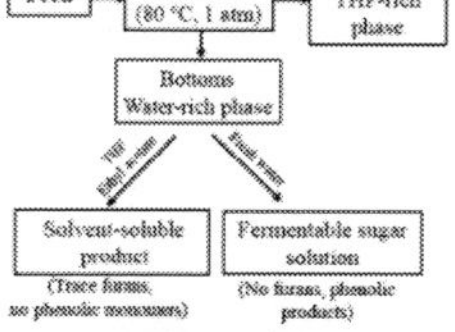

Distillation Fraction	THF recovery (%)
THF-rich phase	106.3
Water-rich phase	0.3

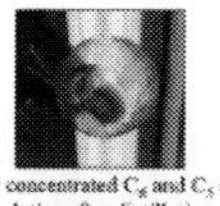

Highly concentrated C_6 and C_5 sugar solution after distillation

Technoeconomic Analysis

- MFSP for GVL and THF processes were $3.33/gal and $2.80/gal ethanol, respectively
- Improvement over both enzymatic hydrolysis and GVL processes

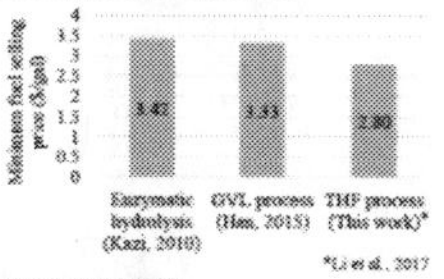

*Li et al., 2017

Conclusions

- Solvent liquefaction with THF/water is a novel non-enzymatic strategy for cellulosic sugar production that achieves high yields of fermentable sugar rapidly
- Delignification in pretreatment enriches biomass with cellulose and loosens up the polymeric structure thus making it more susceptible to deconstruction by solvent and acid catalyst
- Use of THF has economic benefits over current solvent liquefaction processes due to ease of separation

References

Acknowledgements: Prof. Robert C. Brown (Major Professor), Bioeconomy Institute, Iowa Energy Center, Iowa EPSCoR

For more information please attend: Presentation on Thursday, Nov 2nd, 9:40 AM at Room 101D,
Session: Integrated Thermochemical and Biochemical Processing for Renewable Fuels and Chemicals Session

Pericyclic Reactions in Xylose Pyrolysis and Implications for Xylan Pyrolysis

Charles J. McGill and Phillip R. Westmoreland

Department of Chemical and Biomolecular Engineering, North Carolina State University

Xylose as a Model for Xylans

Xylose is a five-carbon monosaccharide. It is found as a constituent of hemicellulose, especially in xylans. Xylose monomers are linked together in the backbone of xylans by 1,4-glycosidic bonds. Xylan also features branching and acetylation at irregular intervals.

β-D-Xylopyranose

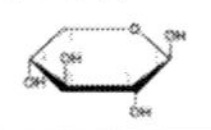

Xylan

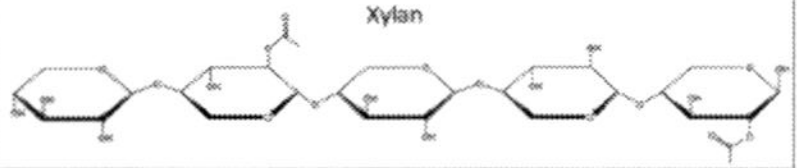

Hemicellulose is a significant component of plant biomass. As such, the chemistry of xylose is significant for the chemistries of hemicellulose and of biomass as a whole. Reaction types found in xylose are applied in this work to xylobiose and 2-acetylxylose as model compounds with similar structures to xylans. The numbering of carbon centers is indicated.

Xylobiose

2-Acetylxylose

Biomass Pyrolysis

Pyrolysis is the process of heating a material in the absence of oxygen or an oxidizer. Pyrolysis of biomass produces water, permanent gases such as carbon dioxide, solid char residue, and condensable vapors. The reactions of pyrolysis are considered to occur in the temperature range of 200-700°C. Both the condensable vapors and the solid char from biomass pyrolysis have applications as fuels.

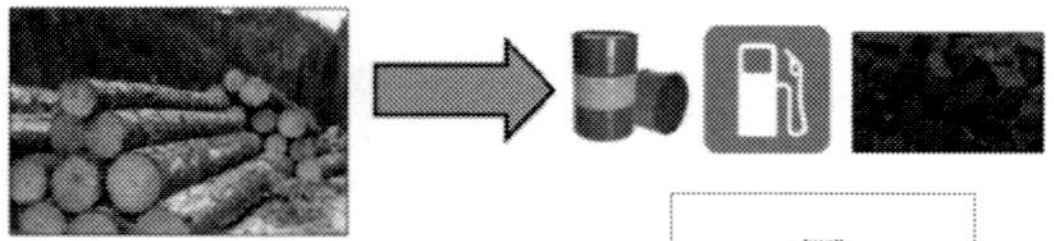

The reactions of pyrolysis have long been studied and typically have been modeled as lumped reactions leading to classes of products[1,2]. It is the goal of this work to develop a mechanism of distinct elementary reaction steps with identified transition states.

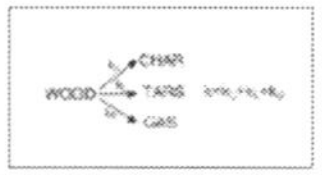

Methods

Calculations of optimized structures and energies were performed using the Gaussian 09 software package[3]. Initial discovery of geometries and reaction transition states were carried out at a B3LYP level of theory with 6-31 G(d,p) orbital models. Structures for xylose and acetylated xylose were further refined using the CBS-QB3 composite method. Activation energy was calculated as the difference in heat of formation at 298 K between the reactant and the transition state.

Reactions of Xylose

Lactol Ring Closure

Energy Barrier (kcal/mol)		
	B3LYP	CBS-QB3
Ring Closing	29.8 – 32.5	31.6 – 34.2
Ring Opening	34.3 – 41.6	37.9 – 46.1

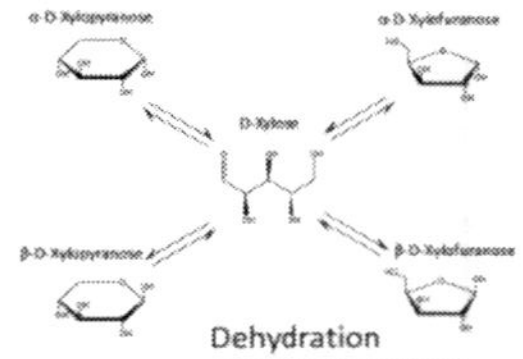

Dehydration

Energy Barrier (kcal/mol)		
	B3LYP	CBS-QB3
H1 to OH2	78.3	83.5
H2 to OH1	70.8	73.0
H3 to OH2	73.6	84.1
H3 to OH4	71.2	76.0

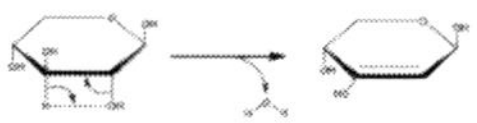

Ether Condensation

Energy Barrier (kcal/mol)		
	B3LYP	CBS-QB3
OH3 to OH1	58.1	62.7
OH2 to OH4	69.5	77.4

Ring Contraction with Dehydration

Energy Barrier (kcal/mol)		
	B3LYP	CBS-QB3
OH1 to OH2	63.0	69.1

Reactions of Xylobiose

Elimination

Energy Barrier (kcal/mol)	
	B3LYP
H3 to bridge-O	60.7
H5 to bridge-O	67.4
H7 to bridge-O	63.7

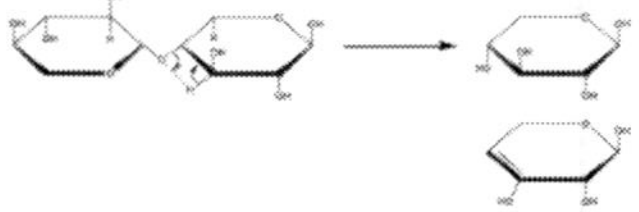

Condensation

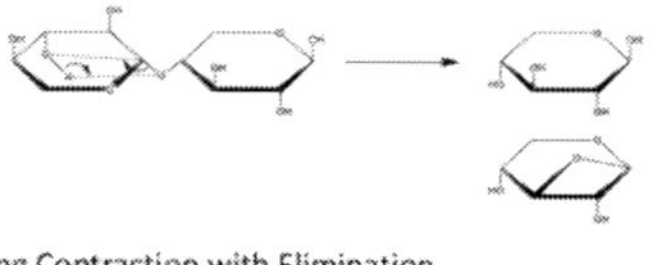

Energy Barrier (kcal/mol)	
	B3LYP
OH2 to bridge-O	63.3
OH8 to bridge-O	52.0

Ring Contraction with Elimination

Energy Barrier (kcal/mol)	
	B3LYP
OH3 to bridge-O	71.3
OH7 to bridge-O	70.5

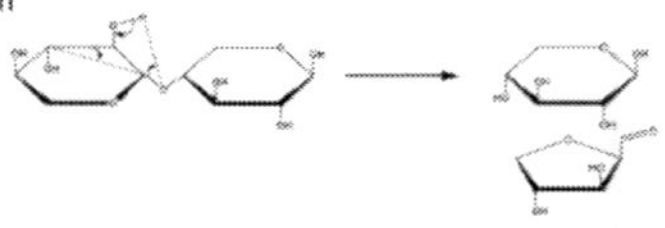

Ring Contraction with Ring Opening

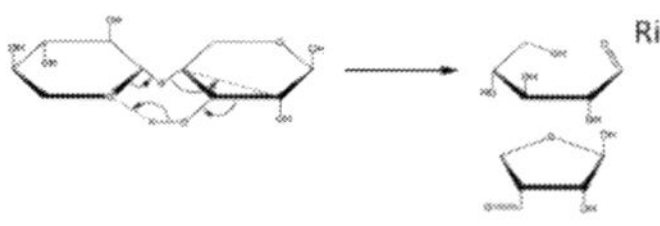

Energy Barrier (kcal/mol)	
	B3LYP
OH3 to O10	66.9

Reactions of 2-Acetylxylose

Elimination

Energy Barrier (kcal/mol)		
	B3LYP	CBS-QB3
H1 to O2	63.5	70.8
H3 to O2	68.2	74.6

Elimination Through Carbonyl

Energy Barrier (kcal/mol)		
	B3LYP	CBS-QB3
H1 to O6	56.3	62.8
H3 to O6	51.3	56.4

Decarboxylation

Energy Barrier (kcal/mol)		
	B3LYP	CBS-QB3
H1 to C7	100.7	107.3
H3 to C7	96.2	102.1

Ring Contraction

Energy Barrier (kcal/mol)		
	B3LYP	CBS-QB3
OH1 to O2	51.5	57.9

Ring Contraction Through Carbonyl

Energy Barrier (kcal/mol)		
	B3LYP	CBS-QB3
OH1 to O6	38.7	46.4

Lactol Ring Closure

Energy Barrier (kcal/mol)		
	B3LYP	CBS-QB3
OH1 to C6	45.7	48.6

Conclusions

1. Isomerization between xylose anomers is accessible with relatively low activation energies.
2. Breaking of the xylobiose glycosidic bond can proceed through several pathways.
3. The lowest energy transition to break the xylobiose glycosidic bond is by a condensation reaction, which unlike the other pathways requires an inversion of the chair isomer to proceed.
4. Elimination of the acetyl group from 2-acetylxylose has a lower energy transition when the carbonyl is involved in the reaction and when removed in a ring contraction rather than a direct elimination.
5. The lactol ring closure with the acetyl group is reversible and can allow the acetyl group to migrate to different positions around the ring.

References

1. Di Blasi, C. Modeling Chemical and Physical Processes of Wood and Biomass Pyrolysis. Prog. Energy Combust. Sci. 2008, 34 (1), 47-90.
2. Dufour, A.; Ouartassi, B.; Bounaceur, R.; Zoulalian, A. Modelling Intra-Particle Phenomena of Biomass Pyrolysis. Chem. Eng. Res. Des. 2011, 89 (10), 2136–2146.
3. Frisch, M.J.; Trucks, G.W.; Schlegel, H.B.; Scuseria, G.E.; Robb, M.A.; Cheeseman, J.R.; Scalmani, G.; Barone, V.; Mennucci, B.; Petersson, G.A.; et al. Gaussian 09, Revision A.1; Gaussian, Inc.: Wallingford, CT, 2009.

Acknowledgements

We gratefully acknowledge support of this research by the U.S. National Science Foundation / CBET under award 1512401, "Establishing Elementary Mechanisms of Biomass and Hemicellulose Torrefaction."

Computational resources:
- NC State University henry2 computing cluster and support.
- SDSC comet supercomputer through the Extreme Science and Engineering Discovery Environment (XSEDE), which is supported by National Science Foundation grant number ACI-1053575.

NC STATE UNIVERSITY

501541 Analyzing the Torrefaction Products of Galactomannan and its Monosaccharide Constituents

Abstract:

Galactomannans are hemicellulose copolymers often found in lignocellulosic biomass and play a role in biomass torrefaction, so measuring their torrefaction products and kinetics is of interest. They are thermally more stable than other hemicellulose polymers such as xylans. While galactomannan's pyrolysis at higher temperature (600°C) has been considered [1], its torrefaction kinetics and its differences versus the torrefaction kinetics of constituent monosaccharides (D-galactose and D-mannose) are not well understood. In the present study, an effort has been put to understand the differences.

For torrefaction experiments, a 0.1 – 0.3 mg sample of galactomannan (Alfa Aesar) or D-galactose (Acros Organics) or D-mannose (Acros Organics) is dried at 105°C for 15 mins before being flash-torrefied (Pyroprobe Model 5000, CDS) at 300°C and held for 30 s. Product analysis is by two-dimensional gas chromatography followed by time-of-flight mass spectrometry (Pegasus 4D, LECO). Products are considered to be identified if they have greater than 80% similarity with the library spectra and their S/N ratio is greater than 100.

Somewhat surprisingly, no 5-hydroxymethylfurfural (5-HMF) or no anhydrosugars were detected from the torrefaction of galactomannan. Both of them are reported to be major pyrolysis products at 600°C [1]. Other major products include water, carbon dioxide, glycoladehyde, acetaldehyde, acetol, glycidol, furanmethanol, furfural, 2-(5H)-furanone, 1,2-cyclopentadione, 2-hydroxy-gamma-butyrolactone, dihydro-4-hydroxy-2-(5H)-furanone.

By comparison from D-mannose and D-galactose torrefaction, strong peaks for furfural, 5-hydroxymethylfurfural, anhydropyranose and anhydrofuranose sugars are detected along with lighter compounds (molecular weight <=74). 3-Hydroxy-2,3-dihydromaltol is also identified as a product, which is not found in galactomannan torrefaction.

[1] X. Zhou, W. Li, R. Mabon, L.J. Broadbelt, L. J., "A critical review on hemicellulose pyrolysis," *Energy Tech.*, 2016, 4, 1 –29.

501573 Thermal Deconstruction Opens Biomass for Acid Hydrolysis to Monosaccharides

Abstract:

We have explored thermal deconstruction, which involves rapidly heating then cooling biomass in an inert environment before complete pyrolysis can occur. While we produce some bio-oil and light gases, solid or liquid phase reactions greatly modify our biomass. As shown through our microscopy and chemical analysis, thermal deconstruction cracks the biopolymers and loosens their lignocellulosic structure. This change allows molecules to diffuse more readily into and out of the biomass. Additionally, we have proven that as the biopolymers fragment into oligomers they become more susceptible to further depolymerization reactions. In particular, when cellulose is thermally deconstructed it rapidly forms anhydro-oligosaccharidesâ€"oligosaccharides where the reducing end has dehydrated. We can hydrolyze these oligomers to glucose under more mild conditions than required for untreated cellulose. Dilute sulfuric acid catalyzes hydrolysisâ€"primarily in cellulose and hemicellulose oligomersâ€"producing glucose and xylose, which can be fermented to ethanol or other products. Our process allows for higher monosaccharide yields than those obtainable from fast pyrolysis.

502236 Conversion of Carboxylic Acids to Linear Olefins By Combined Hydrogenation/Dehydration Reactions

Abstract:

Olefins are important building blocks for the synthesis of higher value chemicals in the petrochemical industry and they are typically produced from fossil fuels. As an alternative to fossil fuel-based chemicals, biomass-derived carboxylic acids could serve as a viable replacement feed for olefin production. However, the production of olefins from carboxylic acids in one step remains a challenge due to the stability of carboxylic acids and the tendency of olefins to over-hydrogenate during hydrogenation, leading to alkanes.

In the present study, we have studied the one-pot catalytic conversion of linear aliphatic carboxylic acids to linear olefins via tandem hydrogenation/dehydration reactions. Hexanoic acid was converted to a mixture of hexenes over Cu nanoparticles supported on commercial silica-alumina (Siral 70) in a continuous flow reactor. The combination of the hydrogenation properties of earth-abundant Cu and the acid sites on silica-alumina allows hydrogenation and dehydration to occur simultaneously, a process that normally calls for different reaction conditions. At 483 K and 5 bar H_2 gauge pressure, we obtained 98.5% conversion of hexanoic acid, with a molar product distribution of 88.9% hexenes and 11.1% hexane. Remarkably, we observed a brusque selectivity switch to 99.9% hexane at 100% conversion. We hypothesize that the presence of a small amount of hexanoic acid on the surface of the catalyst prevents overhydrogenation of the hexenes. Current efforts are targeted at reaction kinetic studies as well as surface studies using FT-IR spectroscopy to ascertain the nature of this selectivity switch. Process runs using carboxylic acids produced from biomass fermentation will also be carried out to ensure process compatibility.

502241 Protection of Carbohydrates during Biomass Deconstruction Using Formaldehyde

Abstract:

Protection of Carbohydrates during Biomass Deconstruction using Formaldehyde

Biomass-derived carbohydrates are important platform molecules for the production of renewable fuels and chemicals. The production of carbohydrates from lignocellulosic biomass requires the extraction of lignin and the cleavage of ether bonds in hemicellulose (mostly xylan) and cellulose chains while minimizing further degradation of the resulting carbohydrates.(*1*) Current methods lead to incomplete biomass depolymerization (producing only polysaccharides) and high process costs due to mineral acid recovery and enzyme production.(*2*) Lowering acid use to improve process economics requires the use of higher temperatures and generally leads to significant sugar degradation and low yields, which is why these strategies have generally been difficult to implement.

Figure 1. (A) Xylose conversion to diformylxylose and furfural during biomass pretreatment (B) Glucose conversion to diformylglucose isomers and 5-hydroxymethylfurfural (5-HMF) during cellulose acid depolymerization.

Recently, we have discovered that formaldehyde (FA) could be used to stabilize lignin and facilitate the conversion of extracted lignin to monomers at high yields (up to 97% of theoretical yield).(*3*) This method could also prevent xylan degradation by producing a stable xylose-derived molecule that we refer to as diformylxylose. In the current work, we study the stabilization of carbohydrates by the addition of FA during integrated biomass depolymerization. The low water content and the acidic environment in the biomass pretreatment allow FA to react with xylose, forming diformylxylose (Figure 1-A) at yields above 90% and minimizing xylose degradation into furfural. In comparison, reactions without formaldehyde lead to almost full xylose degradation into furfural with only 16% xylose recovery. Diformylxylose could be used as is or converted back to xylose at high yields in aqueous environments. A similar process is observed with glucose during cellulose acid depolymerization, forming diformylglucose (DG). The presence of FA led to the formation of two DG isomers from glucose by forming 1,3-dioxolane and 1,3-dioxane structures (Figure 1-B). As with DX, these structures stabilize glucose after depolymerization at conditions that were previously unfavorable due to carbohydrate degradation. Future efforts include the characterization of protected sugars and the removal of their protective groups.

References:

1. J. S. Luterbacher *et al.*, Nonenzymatic Sugar Production from Biomass Using Biomass-Derived γ-Valerolactone. *Science*. **343**, 277–280 (2014).

2. L. Shuai, Y. M. Questell-Santiago, J. S. Luterbacher, A mild biomass pretreatment using γ-valerolactone for concentrated sugar production. *Green Chem.* **18**, 937–943 (2016).

3. L. Shuai *et al.*, Formaldehyde stabilization facilitates lignin monomer production during biomass depolymerization. *Science*. **354**, 329–333 (2016).

492680 A Microkinetic Model for the Catalytic Upgrading of the Bio-Oil Model Compound Acetic Acid

Abstract:

In previous work, we developed mechanistic models of pyrolysis of the three constituents of biomass â€" cellulose, hemicellulose, and lignin â€" capable of capturing experimental yields and trends of primary products present in bio-oil. The poor quality of this liquid bio-oil can be improved by removal of oxygenated products, typically through catalyzed conversion of the pyrolysis product vapors before condensation into specialty chemicals and fuels. In this work, we developed a microkinetic model for the acid-catalyzed upgrading of a model compound, acetic acid, over a common zeolite, HZSM-5. Using a reaction family approach, we developed plausible elementary reaction families for acid catalysis, which involve the formation of carbenium and oxonium ions. We then utilized automatic network generation to construct a reaction mechanism, considering all possible reactions given the reaction families that were defined. The rate coefficients were calculated via the Arrhenius equation, where the activation energy was related to the heat of reaction by the Evans-Polanyi equation. The heat of reaction on the zeolite surface included the heat of reaction in the gas phase, heats of adsorption for the neutral species, and stabilization energies for the ionic species. The heat of reaction in the gas phase was calculated from the heats of formation of all the reacting species. These heats of formation were, in large part, generated from a group additivity scheme, including values for groups involving C+ and O+. In some cases, these values were missing from the literature, so high-level quantum chemical calculations were used to provide values for regression of the missing groups. The reaction mechanism was coupled with reactor design equations to predict product yields and to explore the effect of various operating parameters (e.g., temperature, acidity) on the product distribution.

497152 Upgrading Fast-Hydropyrolysis Products of Cellulose to Higher Molecular Weight Products Using Systems-Level Molecular Mapping

Abstract:

Many of todayâ€™s chemicals are derived from fossil resources, which are not only limited, but have a detrimental environmental impact particularly increasing greenhouse gas emission. Producing chemicals in a fossil fuel deprived future requires systematic utilization of renewable carbon sources such as biomass. Biorefinery concept entails the systematic design of processes that convert biomass to chemicals and fuels. Currently, most biorefinery designs do not thoroughly consider the underlying biomass conversion reaction systems and only focus on existing processes, which do not produce all chemicals and fuels that come from traditional refineries.

Toward developing an integrated biorefinery, we develop a holistic and systematic approach to biorefinery design by fully considering the underlying biomass conversion reaction systems. Our approach considers the biomass-to-chemicals landscape, which mainly consists of primary and subsequent processes. Using automated reaction network generator, we exhaustively define the possible search space of the reaction system. We then construct a reaction-separator superstructure and determine the optimal reaction route through optimization. In order to select thermodynamically feasible reaction route, we also consider thermodynamic parameters in our approach.

To illustrate our approach, we present a case study on upgrading fast-hydropyrolysis of cellulose to higher molecular weight molecules. Levoglucosan and glycol aldehyde are the major components in fast-hydropyrolysis of cellulose. Identifying the optimal reaction route to convert these molecules to higher molecular weight compounds are crucial in order to improve their fuel qualities. In this study, we identified several promising reaction routes to convert the vapor to higher molecular weight compounds and showed that key intermediates are required in order to allow carbon coupling between glycol aldehyde and levoglucosan.

503259 Converting Lignin to Value-Added Products with Copper-Doped Catalysts in Supercritical Methanol

Abstract:

The rapid worldwide increase in the consumption of fuels and chemicals has led to concern over the depletion of non-renewable resources and the environmental impact of their processing and utilization. Lignin, an abundant plant cell wall polymer that is currently treated as waste, is being considered as a renewable feedstock for the production of chemicals due to its high aromaticity. However, effectively converting lignin into value-added products can be challenging due to its high structural heterogeneity and recalcitrance. The copper-doped porous metal oxides (CuPMO) catalyst has been shown to selectively disassemble lignin into aliphatic or aromatic products, which can potentially be used as fuels or platform compounds for chemical synthesis. In order to better control the lignin disassembly to selective products and improve yields for future catalyst design, we focus on understanding the mechanisms and kinetics of lignin and lignin model polymers disassembly over CuPMO via a suite of novel analytical methods, which involves using chromatography, mass spectrometry, and nuclear magnetic resonance.

497657 Cleavage of Lignin Model Polymers with Copper-Doped Catalysts in Supercritical Methanol

Abstract:

The rapid worldwide increase in the consumption of fuels and chemicals has led to concern over the depletion of non-renewable resources and the environmental impact of their processing and utilization. Lignin, an abundant plant cell wall polymer that is currently treated as waste, is being considered as a renewable feedstock for the production of chemicals due to its high aromaticity. In our previous work, the Cu-doped porous metal oxides (CuPMO) catalyst has been shown to convert lignin into aromatic compounds, which can potentially be used as platform compounds for chemical synthesis. However, secondary reactions (e.g., arene hydrogenation) to lead to a wide distribution of products limited the usefulness of lignin depolymerization. In order to further understand the mechanisms and kinetics of lignin depolymerization over CuPMO catalyst and better control the undesired reactions, we synthesized Î²-O-4 linked lignin model polymers, depolymerizing those polymers with CuPMO catalyst and supercritical methanol at 300 °C in as a time-series of 1-9 h. The lignin model polymer and the organic liquids produced have been characterized using chromatography, mass spectrometry, and nuclear magnetic resonance. This suite of analytical methods was conducted in an effort to establish the reaction networks occurring during lignin depolymerization and understand the effect lignin molecular weight and topology has on its depolymerization.

510403 Overcoming the Challenges of Using Corn Stover As Feedstock in Autothermal Pyrolysis"

Abstract:

Corn stover biomass is abundant across the Midwestern United States and is considered a potentially promising feedstock from which biorenewable fuels and chemicals could be produced. However, significant challenges to economically and efficiently convert the lignocellulose fractions into valuable products remain. Fast pyrolysis is an endothermic process capable of thermally deconstructing biomass into bio-oil vapors, non-condensable gases (NCG's), and solid biochar at an elevated temperature (500°C). Autothermal operation differs from traditional pyrolysis with the addition of a small amount of air introduced in the reactor to generate sufficient exothermic energy to drive pyrolytic reactions and achieve a neutral energy balance. Pilot scale testing in a bubbling fluidized bed reactor resulted in low equivalence ratios (ER) during autothermal operation that ranged from 8-11% of the theoretical air requirements for stoichiometric combustion; these ER's are almost half of those compared to autothermal gasifiers. Significant process intensification was reached when air-blown autothermal operation was achieved using 100% air as the fluidizing gas and a maximum biomass feed of 20 kg hr^{-1} was obtained. Bio-oil yields greater than 50 wt. % were achieved along with biochar and NCG's yields around 20 and 30 wt. % respectively. Six bio-oil product streams were collected using a fractional recovery system consisting of three alternating pairs of condensers and electrostatic precipitators; this allowed bio-oil compounds across a range of molecular weights to be collected based on the compound's dew point. Early technical challenges of processing corn stover involved the accumulation of biochar in the reactor and degradation of bio-oil heavy ends in the first stage fraction (SF1). These were overcome by adjusting the fluidization gas velocity to promote better char particle attrition and elutriation from the reactor as well implementing a new SF1 quench tower to facilitate the condensation and collection of viscous bio-oil products. Autothermal process intensification represents a significant advancement in pyrolysis technology and removes the heat transfer bottleneck traditionally associated with thermochemical conversion. Under these conditions, the reactor's footprint was reduced by 40% while simultaneously increasing biomass throughput by a factor of four. This advancement is expected to significantly improve the scalability and process economics of corn stover pyrolysis systems.

510404 Development of a Kinetics Model for Autothermal Pyrolysis in aÂ Fluidized Bed Reactor

Abstract:

Autothermal pyrolysis is achieved by reacting a small amount of oxygen (equivalence ratios of 0.06-0.10) with biomass in a fluidized bed reactor, which partially combusts the products of pyrolysis to overcome the endothermic enthalpy of pyrolysis. Experiments performed at Iowa State have shown that autothermal pyrolysis produces no significant loss in bio-oil yield or quality as compared to traditional non-oxidative pyrolysis. Current experimental data shows that half the heat comes from oxidation of non-condensable gases (NCGs) such as carbon monoxide, methane, ethane, and ethylene and an additional one-quarter comes from oxidation of char. However, currently available kinetic mechanisms for gas-phase combustion of NCGs do not predict significant oxidation at pyrolysis temperatures (400 - 600ËšC). On the other hand, many of the volatile organic compounds that subsequently condense to bio-oil are predicted to readily oxidize at these temperatures, thus prompting a closer examination of the gas-phase kinetics of autothermal pyrolysis. The role of gas-phase oxidative reactions in providing the enthalpy of pyrolysis during autothermal pyrolysis are explored in this study using both experimental and computational methods. A fluidized bed operated at temperatures between 400-600ËšC was used to estimate the oxidation rates of bio-oil model compounds at equivalence ratios of 0.06. Gas analysis allowed product distributions to be determined. These results were used for modeling oxidative reactions during autothermal pyrolysis.

AUTHOR INDEX

AUTHOR INDEX